ETHEREUM

The Comp ete Step by Step Guide to Blockchain
Technology

(The Comprehensive Guide to Funding in
Ethereum & Blockchain Cryptocurrency)

Danny Blackman

Published By Bella Frost

Danny Blackman

Ethereum: The Complete Step by Step Guide to Blockchain Technology (The Comprehensive Guide to Funding in Ethereum & Blockchain Cryptocurrency)

ISBN 978-1-77485-264-4

Legal & Disclaimer

Table of Contents

Introduction

Imagine making a small investment and then watching as your investment grows to millions. Does that sound feasible? With Ethereum it is.

In the past, Bitcoin was the unchallenged the king of the crypto-jungle. However, the throne of Bitcoin is now in doubt and is under threat by Ethereum as the second biggest cryptocurrency according to market cap and one with the ambition to transform into a global super computer. Since its inception it has seen the value of Ethereum has increased by more than 45,800%, transforming everyday Joes into millionaires overnight and billionaires. It is a good thing that you can make your investment an income with the help of Ethereum.

This book you're going to get all the information you must be aware of about Ethereum. Learn about the basics of what Ethereum is and how it functions and how you can make Ethereum as well as how you can keep your Ether safe and also the best ways you can invest your money in Ethereum. After reading this book, you'll be ready to venture into the cryptocurrency market and begin investing into Ethereum without hesitation.

Are you eager to learn about a brand new technology that could transform the world, and help you earn money? Let's take a dive.

Chapter 1: What Is Ethereum? Ethereum Differentiates From Bitcoin

Anyone who is thinking of investing or trading with Ethereum is likely to have seen plenty of hype surrounding Bitcoin but not as much with Ethereum. Both currencies are expected to be the new wave for our economy. Because of this, thousands of dollars of money are invested into both currencies at record rates and there is no indication that either one of them will be slowing down anytime soon.

Both depend both on Blockchain Technology to operate, they are not the same. That's the only place they have a lot in common. One of the primary distinctions between the two is their primary reason in forming. In the last chapter we discussed the most fundamental distinctions in the use for Smart Contracts, but those distinctions do not end there.

Many people consider Bitcoin as a digital currency. It is a way to purchase goods or services, or to transfer money to others, or even to store it as you would with any other physical currency. The major distinction between Bitcoin and conventional currency is due to the absence of official regulations that regulate the value of Bitcoin. The worth of a Bitcoin is not determined by an authority higher than itself that decides its value, but rather through the concept of supply and demand. The more people who want Bitcoin more, the greater the value.

Nowadays, however there are people who are buying Bitcoin and then converting them into "tokens," which are utilized by a variety of businesses when they launch the initial Coin Offering (ICO). This way you can use your tokens for investing in any business they wish, in the event that they have a cryptocurrency available. The tokens function similar to buying shares in a company through its Initial Public Offering (IPO). The price fluctuation of these tokens however they are governed by a different market from that of Bitcoin.

Many people choose to purchase Bitcoin due to a variety of reasons. Some people want another method to keep their money. It does not need a great deal of regulation and oversight like of the other banks. Others may choose to purchase Bitcoin as an investment tool with the goal to hold it for a couple of months or even a few years and waiting for the price to increase significantly, thus making them money. Others may also utilize Bitcoin to gain access to entering the first floor in an ICO of an investment tool. Because a spot in these businesses can only be obtained through the purchase of a token it is the only method of gaining the chance to make profit as they grow.

Ethereum is, contrary to what it appears it offers a completely new way to earn money that Bitcoin can't even come any close to. Many predict that in the next few months, Ethereum is going to surpass Bitcoin as the leader in cryptocurrency in the current economic system.

Since Bitcoin has been the first cryptocurrency to be created that has increased the number of people who are familiar with it. In actual it is the case that all the other "coins" are evaluated by their relation to Bitcoin. Take, for instance, Litecoin which is a lot like Bitcoin in that it has the same features like Bitcoin (it is its own marketplace, has its own name and has it has its own value) however, it's not valued at a level that is comparable to Bitcoin. Although Bitcoin is not worth $15,000+ as at this point however, the value of Litecoin is below $300.00. At present, Litecoin has a very well-known status in the world of digital currencies but is far from becoming market leader, even though it has similar features and characteristics like Bitcoin.

However, Ethereum is unique, not only in the way it functions, but more so in its technology. It's not simply another cryptocurrency. Its "Ether" can be purchased and sold in the same manner like Bitcoin It can also be employed as a way to fund ICOs but its potential goes well beyond the capabilities of. Ethereum remains the same way that Bitcoin has stopped; it's more than just a cryptocurrency, its technology enables its users to develop new programs that perform more than a trade or purchase. This means that it makes Ethereum more of a tool to be used in the future. The ability to develop Smart Contracts means that Ethereum is able to adapt to a larger array of innovative capabilities that might not have even been thought of.

Ethereum has also received a good number of supporters to the Enterprise Ethereum Alliance. It is a group made up of Fortune 500 companies that have already come together to discuss new ideas that could be implemented to the advanced technology. With the advent of Smart Contracts, more and more companies are likely to be looking to make use of this innovative technology. When the transaction details are encoded in the blockchain a lot additional functions can be executed without the need for a intermediary to ensure that everything is completed correctly. Even the most complex tasks can be completed quickly and efficiently saving both parties a lot of cash without the need to shell huge transaction charges for each and every operation in the world.

Although there are a lot of technical factors that differentiate Bitcoin from Ethereum The main distinction is their primary purpose. Bitcoin's primary function has been to be a substitute for money. It is an instrument of payment as well as storage of value. Its primary focus is to allow for users to build peer-to peer contracts and apps that be used to address a broad range of potential. Both are digital currencies and both utilize the latest Blockchain Technology However, Ether is not utilized to substitute for payment, but more of an opportunity to put more power into the users' hands by enabling them to create applications that are decentralized and more appropriate to the requirements of their businesses. Ethereum is, therefore, an advanced network that was created with Blockchain Technology.

Chapter 2: How To Mine Ethereum

This chapter will explore the intricacies of Ethereum mining the process, its complexities and how to begin. The most effective way to begin is to describe the general mining process. You may have seen words such as Ethereum mining or Bitcoin mining But do you understand what it means to be a mining?

The Mining Process: Understanding the Mining Process

In essence mining refers to the discovery of something valuable or precious hidden beneath the earth. Mining for digital currencies works exactly the same way as physical mining. When it comes to cryptocurrency mining, it's lots of computing power as well as time, and participation.

In the first chapter we spoke about networks while we were working out the definition of the blockchain. When a person joins the network and is able to perform various computations efficiently, they receive rewards through cryptocurrency.

Mining is finding the rewards you want by taking part in solving mathematical issues and calculations. If you're a mining professional then you make use of the computer's hardware to do these calculations for you , so you can earn money.

You might be wondering why the people who issue these digital currencies can't perform this task on their own. Why do they require other individuals to handle the mining? Issuers require the help of the participants since they can't do this work on their own. They don't have the ability to manage these tasks by themselves, so they solicit help from people all across the globe to step to help.

The description might sound simple but mining is an arduous procedure that requires an enormous amount of time. If you are planning to become mining, you will need to use your computer to carry these duties. Your computer will always be looking through blocks to find solutions for equations and permutations.

If you're able to make these calculations, you send your solution in writing to your issuer. The issuer will go through your answers and award the user with digital coins. Similar to mining for precious metals and gems it is possible to mine for digital currencies using an enormous amount of computing power.

Mining is crucial as it helps increase the quantity of coins available on the market. This allows trading to occur. The trading process is made easier once there are sufficient units of cryptocurrency circulating on the market.

Mining Ethereum

Mining Ethereum is more than just bringing additional Ethers in circulation. It's also about protecting the network that generates blocks with the help of the different transactions. In mining you need to consider the Ether unit is crucial as it is the source of energy that fuels an Ethereum network. It is a catalyst for change and growth as people constantly search for ways they can build diverse platforms that work with it. This method encourages creativity and expansion.

After you have mined the Ether after you have mined it, it's up to you to choose what you'd like to make of it. You may trade it in or sell it in exchange for money. There are more than 18 million Ethers issued each year and the figure continues to grow dramatically each year.

If a user is able to complete the transaction, they are required to provide an "proof that they have done their job" to be paid. Every block must be a record of work in order that people who issue the money know who did the task. The proof of work system is an process that ensures that miners are legitimate rather than hackers looking to disrupt the system. Miners must demonstrate that they are legitimate by dedicating the computational power required to solve the mathematical problem.

The process isn't simple since there could be many users. This method (or process) assists in determining if the work proof is genuine. This algorithm for validation is called Esthash. The way it works is that each task has its own level of difficulty, which the system determines. The system regulates the pressure according to the amount of time that a miner is required to accomplish a task. By manipulating the challenge it will determine how long a person must spend to finish an issue. While mining it, the difficulty level is adjusted gradually, and there is around the release of one new block every twelve seconds.

If you're looking to explore mining, you may begin at home, but you'll must be knowledgeable about commands and scripts. Although it might appear as if it's a difficult procedure, the trick is breaking it into small steps. It also requires plenty of electricity which means you can expect to pay high electricity costs. However, you will make money by selling your Ethers. There are calculators on the internet to help you figure out the amount of money you've earned after subtracting your expenses from your income.

The Mining Process

The initial step of Ethereum mining is to purchase an electronic device. Purchase an graphic card (GPU) with an RAM (Random Access Memory) that is greater than 2GB. A tip is that a GPU will be superior to an CPU (Central Processing Unit) since it's more efficient in this kind of mining. Alternatives include Nivida as well as AMD cards. You'll also need to have ample space on your system for your computer to host the program and the various applications.

Once we have a clear idea of the requirements to do, let's take a examine the process in itself:

Download Geth Download Geth Geth is a program that will connect your account to Ethereum network, allowing you to connect with issuers. It keeps you up to current with the happenings on the Ethereum network. It is important to become familiar with how it functions prior to beginning.

Download the zip file It's a zip file. Geth generally comes in an unzip file, you need to extract it and then save it to the hard disk. It is recommended to keep it in Drive C.

Find CMD (command prompt) Run the program.

Find Goh in your personal computer.

Account creation Account creation "geth account created" and hit Enter to change it so that it changes to 'C:>geth Account New.'

Create a password - Note the password down and then save it. Enter it into your account and make a new account.

Download the blockchain, and then connect to the network. In your terminal type in 'geth --rpc' and hit Enter. This will allow you to establish a connection to the network. This can take a considerable time. Be patient and allow it to be completed before you begin mining.

Download mining software. It can help your graphics card connect on the network. The software you could make use of is called Ethminer. A lot of people have used it successfully. Install the program that you plan to make use of.

Find the latest software - locate Ethminer or the program you prefer on your personal computer.

Follow the next steps: Start typing "cd prog" and then press the tab. The following will appear "C:>cd prog.' Press Tab to display the following: 'C:/> "Program Files". After that you can press Enter to show 'C:Program files>.'

In the software, type "cd cpp" then press Tab and Enter - You should see 'C:\Program Files\cpp-ethereum>.'

Enter "ethminer G" and hit Enter to begin the process. You will get a graph that is called"the DAG (Directed Acyclic graph). Also, ensure that you have sufficient space on your computer to facilitate this procedure.

Utilizing the CPU - You can still mine using your computer by entering "Ethmining" while pressing the button.

Do not expect mining process to be simple. It's a lengthy procedure that involves several going back and forth. The profit will be small for the first few months due to the initial and operating costs are very high. But, over time it will be possible making money by mining.

Following the steps described in this chapter, you are able to begin your mining process using the Ethereum network. The process could earn you some cash however make sure that you've weighed the risks and advantages to do not lose money over the long term.

In the next chapter, you'll be taught how to create an account and make transactions with exchanges.

Chapter 3: The Way To Buy Ether

In this chapter, you're going to be taught different ways to purchase Ether and also how you can use all of them.

The most simple and straightforward method of purchasing Ether is through an exchange for cryptocurrency. These are platforms online which facilitate trading of cryptocurrencies among users. Because of the popularity of Ethereum and massive market capitalization There are a variety of cryptocurrency exchanges that accept Ether that means you can choose from a variety of options regardless of where you live. Although different exchanges for cryptocurrency use slightly different methods however, the procedure of purchasing Ether via an exchange that includes the following:

The first step to accomplish is to set up an account with the exchange of preference. This will require you to provide certain details about yourself, including your name, address as well as your email address. When you sign up for an account it is essential to verify that the exchange supports for your country as well as the fiat currency you plan to make purchases with. Ether.

After you've signed-up for an account, many exchanges will require more details before you are able to make withdrawals or deposits. Most of the time they'll ask for your government-issued ID as well as a picture of you as well as identification of your address. This is to make sure that the transaction is n conformity to Anti-Money Laundering (AML) and Know Your Customer (KYC) laws.

After you've confirmed your identity and address, you are now able to select the method of deposit you prefer. Different exchanges have various deposit methods, and you must verify this prior to registering for an account. This information can be located on the website of the exchange together with the charges for each option. The most commonly used deposit options are the wire transfer, PayPal payments, SEPA and credit and debit cards.

After you have identified the deposit method that is best for you, then you are able to proceed to deposit your currency in an exchange service. Euros and Dollars are accepted in almost every exchange and a number of other exchanges accept other major fiat currencies like Sterling Pounds Canadian Dollars Japanese Yen, and Chinese Yuan. The deposit could take between a few hours to several days to be reflected in your account on exchange platforms depending on the deposit method.

When your funds are reflected in your account on exchange platforms and you're ready to purchase Ether. The procedure varies based the exchange you choose however, most exchanges attempt to make it simple and easy for novices. After you've received the Ether on your exchange platform account it is recommended to transfer them to a bank account whose keys you manage. Don't leave them to the exchange site.

The most well-known Cryptocurrency Exchanges where You Can Purchase Ether

Coinbase

It is the world's most well-known cryptocurrency exchange and is a great option for those looking to purchase Ether in Canada, the USA, Canada, the UK, Europe and Singapore. Alongside Ether, Coinbase supports several other cryptocurrency. Coinbase has been operating for more than six years, and has earned itself the status of among the top reliable and trusted exchanges. Making purchases of Ether via Coinbase is simple and straightforward. If you're an skilled user, you could make use of Coinbase's GDAX which has a more sophisticated range of options. Coinbase lets users deposit money (fiat currency) via banks transfers, SEPA, and debit and credit cards. If you choose to pay with credit/debit cards then you'll need to face a smaller limit. Although it is the most well-known crypto exchange Coinbase can only be found in 32 nations. So, it is important to confirm that your country is supported prior to joining. It is also important to know that Coinbase doesn't sell Ethereum Classic.

CEX.io

Another popular exchange that's been in operation for quite a while. It's associated with Fincen and offers brokerage services as well as being a cryptocurrency exchange. Cex.io has begun to support Ethereum in the year 2016 following the fact that they have brought their cloud mining services up to the end of. Contrary to Coinbase, Cex.io is available all over the world. You can transfer funds through Cex.io by wire transfer either through SEPA or credit/debit cards. Verification the possibility of purchasing an additional amount of Ether by credit card. Its Cex.io website is simple and user-friendly. This makes the process of buying quite simple even for novices. On the other hand the costs paid by Cex.io are a little more expensive. But, it's an option for those who reside in countries that aren't accepted by Coinbase.

Coinmama

Established in 2013, Coinmama is an exchange for crypto as well as a brokerage company that accepts Bitcoin along with Ethereum. One of the biggest advantages of Coinmama is the ability to purchase the equivalent of $125 in Ether without having to be verified. It is possible to visit their website, sign-up, and make your purchase in just 20 minutes. Coinmama will only allow you to purchase Ethereum by debit or credit card. The customer service they provide is good, and the service is accessible worldwide. However, their costs are somewhat excessive.

BitPanda

Formerly called Coinimal, BitPanda is an Austrian cryptocurrency broker that permits individuals in the Eurozone to purchase as well as sell Ether. It was established in 2014 and the platform has seen a surge of attention from users across Europe. BitPanda permits you to pay for Ether via SEPA or credit card Skrill and a variety of other payment options that are widely used across Europe. The fees are quite affordable, and their site is extremely fast and secure.

Gemini

The company was established in the year 2015 by the Winklevoss twins Gemini was founded in 2015 by the twins Winklevoss, New York based cryptocurrency exchange has seen rapid growth in its popularity. Gemini provides its services to customers from North America, Europe and Asia. Gemini is a platform that works in three regions: Europe, North America and Asia. Gemini platform operates as traditional exchanges for forex that allows users to trade directly, with prices determined by the user. Gemini is only able to accept deposits via transfer to a bank account. One of its biggest benefits is that its costs are minimal.

Changelly

It was founded in the year 2016 Changelly was founded in 2016, and is a new player in the market. Although it hasn't been for a long time the company has been gaining a lot of attention. One of the most appealing advantages of Changelly is that it permits users to exchange one cryptocurrency to another. This makes the platform an excellent option for those who already own an amount of Bitcoin that you wish to trade to Ether. While it is possible purchase Ethereum through Changelly with fiat currency the costs are expensive. If you own Bitcoin that you wish to exchange into Ether The process takes around 30 minutes.

Bitfinex

Bitfinex is an Bitcoin exchange that allows clients to buy and sell cryptocurrency. If you own Bitcoin it is possible to exchange it in exchange for Ether through the platform. Bitfinex doesn't support fiat currency, which means that the only way to acquire Ether via it is to purchase the Bitcoin first. Bitfinex is extremely secure, boasting sophisticated security measures to ensure the security of customer information. The funds of customers are kept in cold wallets to protect against hacking attempts on the internet.

Kraken

It is a different US cryptocurrency exchange that allows users to purchase Ether by using fiat currency to exchange it for other currencies. Kraken is among the first exchanges to have been established, after it was founded in the year 2011. Kraken offers more that Gemini or Coinbase. If you decide to purchase Ether from Kraken with a fiat currency, you can deposit funds via transfers to banks. The only criticism I have about Kraken is that their interface for users is a bit sloppy. The customer service is described as being a little slow.

Purchase Ethereum in a secure way

As I said previously, buying Ether via an exchange online requires you to show proof of identity. Sometimes, however, you may prefer to buy Ether anonymously , for one reason or another. In this scenario you can do this by buying Ether via peer to peer exchange platforms like localethereum.com.

Localethereum.com is an online platform that connects Ethereum customers and vendors from the same geographic area. It can be thought of as an online market as similar to eBay but the product is Ether. Both the seller and buyer are able to agree on a payment option that is most suitable for them both, whether it is transfer via bank wire, PayPal, Skrill, credit card, Bitcoin, or even cash. To pay for its work, localethereum.com charges the seller only a small portion of the transaction. This is the best method to buy Ether in a secure manner. However, you will have to be cautious. For example, if you decide to meet with the seller in exchange for Ether in exchange for cash, it is important to consider your security.

Alternately, you can buy Bitcoin and then trade your Bitcoin for ether via shapeshift.io. This exchange lets you exchange different cryptocurrency without the need to sign up for an account on the platform. But, you are not able to purchase the entire amount of Ether via shapeshift.io.

You can also purchase Ether privately by using the Ethereum ATM. They are ATM machines that permit the exchange of cash in exchange for Ether. Ethereum ATM's don't require any identification. All you have to do is select what amount Ether you wish to purchase, then enter your Ethereum wallet's address and then insert money into the ATM. It will then transfer the Ether will be transferred automatically to your account. However, because Ethereum ATMs don't ask for evidence of identity You can only purchase small quantities of Ether at one Ethereum ATM.

How to Select The Best Ethereum Exchange

With the variety of options available and options, choosing the right Ethereum exchange to meet your needs could be quite a challenge. To ensure you're making the right choice here are some things to take into consideration before settling on an exchange platform.

Locations that are supported Certain exchanges such as Coinbase and Gemini only provide their services only in certain geographical areas. Thus, before you sign up for an account on any exchange, you must to confirm that they have services in the area you reside.

Methods for depositing: must confirm that the exchange you choose supports your preferred deposit method. For example, if you wish to make your deposit through PayPal then you can't make use of Gemini. Keep in mind that various deposit methods come with different charges. In general, the quicker and more efficient the deposit method, the more the cost of fees you will be charged.

Supported currency: Different exchanges accept deposits in different currencies. In the majority of cases, this is dependent on the your location. The majority of exchanges will accept USD or Euro deposits. But, if you are planning to deposit any other currency, it's recommended to check first if the currency is accepted at the exchange you prefer. For example, if you want to transfer Japanese Yen to a bank account it is only possible to choose an exchange that is able to accept Japanese Yen for deposits. Also, you are not able to make use of Bitfinex in the event that you plan to purchase Ether by using fiat currency.

Security The purchase of Ether is a financial transaction, therefore you must ensure that you can trust your financial information and funds without a risk that they could be taken by hackers. Go to the website of the exchange to determine the type and security procedures they put in place to guard your funds and personal information.

Support Sometimes, you'll have some issues in setting up your account, or purchasing Ether particularly in the beginning. You need an exchange with a quick and efficient support staff that can quickly assist you in resolving any problems that may occur.

Fees: Different exchanges offer different fees. You should choose one that charges the most affordable fees.

Reputation: Before choosing a particular exchange, it's an excellent idea to research the reputation of the exchange. What is the opinion of other users about it? Look through different forums and review websites to find out the type of reputation that the exchange platform enjoys. If you see a number of complaints from customers, it could be an alarm.

These are just a few of the things you need to be aware of when selecting the right cryptocurrency exchange. But, there are times when it's difficult to find a service that is able to meet all your needs. In these instances you may have to compromise on some of these elements. If, for instance the sole exchange in your area charges excessive fees, you may have no choice but to use it or not buy Ether.

Chapter Summary

This chapter, you've been taught:

The easiest and most basic method of purchasing Ether is through an exchange for cryptocurrency.

Many cryptocurrency exchange platforms request government issued identification, a photograph of yourself , and a evidence of address to be in conformity to Anti-Money Laundering (AML) and Know Your Customer (KYC) laws.

Different exchanges have different deposit methods, therefore it is important to verify the requirements prior to registering for an account on an exchange for cryptocurrency.

When you have received Ether within your account on exchange platforms, it is best to transfer Ether to a bank account whose keys you manage. Don't leave them to the exchange site.

The most popular exchanges for cryptocurrency that allow you to purchase Ether are Coinbase, CEX.io, Coinmama, BitPanda, Changelly, Gemini, Bitfinex and Kraken.

If you are looking to buy Ethereum without revealing your identity you can do it by buying Ether through peer-to-peer exchange platforms, like localethereum.com. You can also buy Bitcoin and then trade your Bitcoin to Ether via shapeshift.io.

You can also buy Ether anonymously at the Ethereum ATM.

When choosing the ideal Ethereum exchange, you must take into consideration factors like accepted location, accepted deposits, currencies that are supported fees, security as well as customer support and reputation.

In the next section, you'll be taught details about Ethereum wallets.

Chapter 4: Ethereum Enterprise Alliance
Ethereum Enterprise Alliance

Fortune 500 companies, blockchain-based startups, and research groups came together in spring 2017 to form a non-profit organisation that's known as the Ethereum enterprise alliance. The businesses that are part of the business are Microsoft as well as Intel. There are at most a hundred sixteen companies who are members of the alliance for enterprise.

Official EEA logo
The alliance was established as an open source project for private blockchains that these businesses use. The blockchains will handle the information stored in blocks for each sector that uses blockchain. This implies that every aspect of entertainment to banking is used by these businesses. The alliance has discovered an answer that is compatible with the ecosystem of ethereum that is used by all users of the ethereum platform and the alliance. The technology the alliance is developing will

not just help to achieve their goals, but is also improving the efficiency of ethereum for all users who use it.

Certain alliance members have revealed that they are launching new blockchain projects based on their blockchain. One of them is a hybrid technology that connects private chains with public chains. The blockchain that is created will be public and open opportunities that were not previously available if it had otherwise been not for this company that put their plans into practice.

The Ethereum blockchain is constantly changing, this means that any information contained in the blockchain of the company is available to the general public to view what the company is up to. Therefore, at all times, Ethereum Enterprise Alliance wants to allow private blockchains and free blockchains. If you make open the blockchains that are closed and businesses' activities, what they're doing will not remain hidden and the public can observe how they operate from inside. There might not be any way that the general public can communicate with the business, but this doesn't mean the public will not be able to be able know what companies are doing. If the general public is able to observe everything, how are big Fortune 500 companies going to to shield themselves from scrutiny when they are confronted with questions about how their funds are employed to provide an improved customer experience.

One of the biggest sectors looking to bridge the gap is the financial firms as they don't want to lose customers. That's why they are looking to join in with the blockchain technology to show their clients that they are eager to participate in the way the future will be and maintain the trust of their clients.

It's more palatable when you can see the company's plans are being implemented. Let's take a look at some companies trying to implement their ideas into practice for users of Ethereum.

JP Morgan Chase started a blockchain that was situated between the blockchains that were private and public in order to enable payments to be made and received. The concept was derived from a regulatory agency that required access to transactions of their business but also keeping the privacy of their customers safe.

The Royal bank in Scotland announced that they had created an instrument to help get settlements cleared out of the ledger distributed. The ledger was based on smart contracts and allow users to record their settlement contracts.

Chapter 5: Roadblocks

Every new product has an initial stage of development. Knowing what's held people back before will speed up your learning process and help get your product on the market sooner. Let's look at some of the roadblocks.

Scalability is a problem that affects the main payment networks due to the fact that there are around 2000 transactions per second. It is essential to modify the limit on block size to ensure that there are more transactions that can be supported per second. There is a possibility at this moment that should Ethereum were to expand in size there would be a risk that only finite number of nodes are able be utilized.

Certain transaction blocks need more than 50% of the hash power on the network to reverse. This is partly due to the security requirements and can be rectified when reverses result in a cost to the initiator.

Stamping is a concern at present, also. In general, in the world of Blockchain, block blocks are created each day. But, making more blocks often causes payment systems to become extremely slow. Therefore, the network operates best with a set amount of blocks created every day. Furthermore, each user has a distinct time date (distributed according to a normal curve) so that there are no nodes less than 20 seconds from one the other, making the communication between nodes deliberately delayed.

There are many other obstacles however, the developers are striving to improve every stage through the entire process of Ethereum and other smart contract. Be patient in waiting for developers to address issues.

Chapter 6: What You'll Need To Know To Start

If you're at this Chapter this assumes you're curious about Ethereum investment and trading in the platform of cryptocurrency. If so you're probably thinking "Great. What can I do to get started?"

Beginning is the simple part. It's also the easiest of everything you read in this guide it is possible to start right now , with just what you have available and without costs.

The first thing you'll have to accomplish to begin is to determine what you'll do earning Ethereum Coins. Are you planning to do Ethereum mining with your computer's resources and resources, to compete with other miners to earn parts of a coin, before you are in a position to put it all together? Are you looking to begin offering services on the internet and then sell your products and skills to make money? Do you have an income stream that is more traditional currencies you would like to spend on buying them via the internet, and then keep the money? Do you wish to do something with both of them?

The answer will depend on your goals, as every decision will require various resources.

Mining is dependent on your computer's processor power. Naturally, older and less powerful computer systems will constantly be when it comes to performance, and could not be able mining efficiently enough to justify the expense and effort you're investing in mining. The applications will be running don't just run on the background on your computer to get on with your day. They need the full potential of all the resources your computer can access and will be operating at maximum capacity. For a majority of Ethereum mining enthusiasts, they usually purchase two or three powerful processors and use them as dedicated mining machines that are designed to be used for nothing else than. Sometimes, they pool their resources and join what's known as an "pool" where others connect their computers and carry out mining tasks in a group and distribute coins equally to all members. Some miners purchase servers from others and use the server as a mining tool when they don't have the resources or time to invest in an equipment for mining. In many ways, even if you don't have the machine specifically designed to mine, as well as the dedicated time and effort to mining mining, the expense could quickly surpass what you receive from it. Sometimes, the cost of mining isn't worth the effort, however, it all depends on other elements.

In all other cases all you require is internet access and a good computer in the mid- to low-end. Much of what you require is an amount of information and is accessible like other web-based applications.

In that regard the next thing you should be focusing on is what kind of wallet system you'd like to utilize.

Wallets

Now, what exactly is the definition of a wallet? It's been mentioned previously in this ebook however, it's not discussed in detail. Does it resemble the traditional wallet, which is utilized to keep money? Yes No.

The traditional wallet is usually simply they are. They store your money easily and securely and also other data. Digital wallets work like traditional wallets however they are primarily focused on managing cryptocurrency. In many ways, they're like the types of wallets that you come across offline which means that the different wallets you can find perform different tasks.

Some wallet programs are created to be specific, and are designed to more specific purposes than others, typically with a focus on security as well as storage. They are ideal for those who wants to concentrate on only one cryptocurrency at a given time and also to ensure that the money they earn is secured from people who like to steal their personal information and hack their accounts in order to steal their funds. Although the majority of their security features are commonplace, many instances, these specialized wallets provide offline services and include customer service numbers that you can contact and IP tracking, print-friendly forms, and the like. they offer.

On the other side of the spectrum, there are wallets made to be more universal. Most wallets offer many options to hold different types of cryptocurrency, not just Ethereum coins. They are typically used by those who are more general in their investing and are particularly useful for those who eventually decide to invest in other cryptocurrencies than Ethereum. The majority of universal wallets can accommodate for the various cryptocurrencies available, including those designed to be an amusing joke. One of the great things with these kinds of wallets is that they allow you to use them to trade with other users who are using similar programs, exchanging one kind of cryptocurrency to another. It's not unusual to see individuals trading lesser-value coins in exchange for Ethereum or other expensive coins.

To start out However, it's usually prudent to begin with the wallet software which is available together with Ethereum through their official website. It's not just extremely specialized and focused on the Ethereum coin and is linked to the company who developed the altcoin, so you'll be able to solve any issues any issues that occur. It is highly advised to explore other wallet programs that are available it is important to conduct your research prior to committing so that you don't fall victim to fraud and have your hard-earned investment taken from you.

However, the next thing to be concerned about is your internet connectivity, as well as protecting your computer from intrusions from outside. This is a given however, since it's generally a good idea to safeguard your system even if you're not associated with the Ethereum trading market however, by making sure you're secure, you'll have less of a risk that you'll have your cryptocurrency stolen. Simple firewall extensions such as anti-virus tools, rootkits, and even anti-malware software can help make sure that no one is able to gain access to your personal information through backdoor software. Always be sure to back up all important data, such as the information on your wallet, information about pools or even your business details and stay alert.

Chapter 7: Ether And Other Crypto-Currencies

Cryptocurrencies have been explained by their creators as well as computer experts, but they remain unknown to the everyone else. Anyone who has looked into crypto-currencies have a good idea of the significance they will be in the near future. In this section, we'll examine what crypto-currencies are, as well as the market on which they trade.

Cryptocurrencies are digital currencies which is transferred with no assistance of central banks. They are secured to protect against fraud or to surveillance from an centralized location. It is possible to view crypto-currencies for entries to a ledger, which can't be altered after the transaction is completed.

Some examples of crypto-currencies comprise however, they are not limited to Bitcoin or Litecoin. Ether. These are only a handful of the crypto-currencies on the trading market for crypto-currencies. The market for crypto-currency serves as a platform for trading and buying of digital currencies from all over the world. It is comparable to the foreign exchange market in which currencies from different nations are traded. The distinction is that digital currencies aren't associated with any particular country, region or organization, nor are they tied to any specific geographical location. The currencies are decentralized.

On the market for cryptocurrency you can exchange with your Bitcoin or Ether to USD and BTC. In this way, you will participate in the cryptocurrency market without taking the dangers associated with mining. Cloud mining, as well as other issues will be discussed later in the chapter.

Like the conventional foreign exchange There are a variety of websites where you can trade your cryptocurrency. Certain of them are more complicated than others. They are the ones employed by experts to trade crypto currencies. They provide access to all the sophisticated tools used for trading. If you're not interested in using the tools, there are websites that are able to trade without the creation of an account.

After you've decided on the type of trading you wish to pursue, the next thing to do is to choose what kind of exchange that you intend to utilize.

There are three kinds of exchanges.

Direct Trading is a type of platform which allows people to trade currencies in a direct manner, from one to another typically in other countries. There is no set foreign exchange rates and the people who take part in the exchange set their own rates.

Trading Platforms: These are platforms on which sellers and buyers are linked to one another. The platform usually charges an amount for each transaction.

Brokers - You can go to these websites and purchase the crypto-currencies direct with the seller. The price is decided at the discretion of the seller. We can compare them with the traditional foreign exchange traders.

Before joining any exchange that is available online it is important to check at a few points:

Restrictions on Country

The digital currency business is open to everyone in the world. However, the exchanges are not as adaptable. Certain of the features that are offered by certain exchanges are accessible only through one specific country. This is particularly important when you plan to purchase any currency. Verify that the services offered by the website are accessible in the region where you are at the moment.

Exchange Rate

It may appear obvious , but it is vital for trading. Some platforms modify their exchange rates frequently. Make sure to check for these changes prior to signing up to any platform.

User Fees

As we've mentioned previously ensure that you are aware of the charges that come from your transactions. For certain platforms, fees to deposit currency are not that high however they increase when you withdraw. Make sure you are aware of these fees prior to signing up to any platform.

Payment Options

More payment methods you have the better. If you discover a payment platform with limitations on payment choices, you'll need to find an alternative platform. Platforms that offer only the option of paying with credit or debit cards could be vulnerable to fraud. You'll need to divulge the details of your credit card which could be dangerous.

Methods of Verification Methods of Verification

Some exchanges need you to confirm your username. Certain exchanges will permit users to remain completely anonymous. The verification process could be longer, but it is definitely worth it. The time needed to confirm your identity prior to making or withdrawing a deposit is a means to stop fraudulent attempts.

There are numerous advantages of trading on the cryptocurrency market as compared to conventional foreign exchange.

The spreads are much smaller.

In the normal market your transactions on your own could cost you funds due to spreads. It is the gap between price of the transaction in the market and bid prices for the currency. If you're using euros or US dollars the spread for the transaction could be substantial. For crypto-currencies, the spread is small and it's as if there were no charges for the transaction.

Margin Trading

It is possible to trade margin-based in the world of crypto-currency. This means that you pay a certain amount and receive the remainder of the money from your friends. The loan money returns to the lending institution in exchange for an interest rate, which lets you make an income from funds that are not yours.

Leverage Trading

It is trading using funds that you don't have. The typical rates for leverage trading on the majority of crypto-currency websites are 1:10. That means, when you

have a dollar, you could trade for 10
dollars even though you don't have the
amount.

Chapter 8: Programming In Ethereum

If you're looking to begin developing your own application for the Ethereum blockchain, the first thing you're going be required do is learn the Solidity programming language that you'll find to be broadly similar to JavaScript. It makes use of the .se and .sol extensions as well as LLL which is a Lisp product. If you've previously worked with Serpent or Python and feel at ease with Solidity.

To make sure that you are able to compile your apps in the quickest way possible it is necessary to have to select the solc compiler that integrates in conjunction with C++. If you're not comfortable working with solc, you could use an alternative browser that works with in-browsers like Cosmo rather, however this chapter assumes that you have chosen solc. After you've completed your work you'll need to utilize the Ethereum Web3.ja API to make sure that you are able to utilize JavaScript in order to link your contact list directly to the application.

This is what will allow you to communicate with your smart contracts no matter where you go, not only when you're using your Ethereum node.

Frameworks

If you're looking to utilize an application framework to build the distributed app you are developing, you'll likely be happy to learn that there is many choices available, at no cost which the community has come up with. They are a great place to begin, and you do not have to worry about creating your own framework.

Truffle The Truffle application is an excellent starting point since it automates all of the general programing steps that are needed for getting a decentralized application running. This allows you to spend less time performing the necessary steps and more time focusing on whatever it is that you have to work on for your application that is certain to be a hit with the world. If you're using Truffle or Truffle, you may want to consider Embark. Both work in tandem and Embark is well-known to be useful in streamlining and developing apps because it can automate a large portion of the testing procedure.

Meteor If you're trying to enhance your stack, Meteor is the choice of many Ethereum developers since it works perfectly with Web3.js API. It also works well with various Web application platforms, and was among the major supporters of the platform in its beginning days.

API: The API the majority of the new Ethereum developers use is one developed by BlockApps.net. It works in the same way as an actual Ethereum node that is great in situations when you're not able to operate a real-time node, but you still want to work on your application. Another alternative is called MetaMask that allows users to access the standard set of Ethereum tools right using a browser on the internet. Another alternative is LightWallet that is a simple option for developers as well as users to connect with decentralized apps while offering different kinds of users with a variety of specialized interfaces.

Making your App

To start the process of creating an application beginning, the first thing you'll have for is to install your personal Ethereum node. It is possible to do this using your command line using Ethereum's Ethereum the node's interface. On your command line, enter the following: bash<(curlhttps://install-geth.ethereum.org).

Once you have completed that step, you will then be asked to start the installation process after you have selected the right operating system and most recent Version of Ethereum CLL. When the installation is finished it wi l be in a position to use Geth through the JavaScript system that can respond to console commands of standard. If you want to develop specific console commands, they will be saved to aid in tracking your progress in the future. After that now, you can begin working. The terminal program to open your Geth console. After you've done this, you will notice a "+" at the corner that indicates everything is running smoothly. To exit, just press EXIT, and then ENTER.

Once you've completed coding your smart contract or application that you will use within an application step you'll have to create a compile. Once everything is compile then you can use the result. To deploy the results all you have to do is pay a fee for gas and then sign a digital contract. After you have signed the contract, you will receive an URL that will direct you to the address of the contract on the blockchain and your ABI of the software.

After you've obtained this ABI you'll be able access the contract or application from any device with internet access. Based on what the application is doing, every time you use it you might be charged the cost of gas.

Testing

If you're building an intelligent contract, it is vital that your if/then statements are written in a way so that there is no in the way of them. This is the perfect time to start loading Truffle since it can instantly create the framework required by Web3.js as well as JavaScript to ensure that the smart contract is working as you intended it to.

When you test the time of transactions and the guarantees you make will play a an important role in determining whether the smart contract or application is actually used by the public. To ensure that the community is aware the community, your app will need to be able to verify as quickly as is possible. The minimum limit currently is 10 seconds, however it is seldom used in non-testing scenarios.

When you are ready to test the contract, you should access your test folder and modify the .js extension to conference.js and be sure to alter any other references to it as well. After this is done, you should start Truffle within the root directory linked with the test files.

Once this is completed, you will have to then start Solc, Solidity and Pip. Prior to opening these you must ensure that your main library is separate from the test library with the help of an environment virtual. Once this is completed you'll then need to launch a console and start a new node client.

Then, you're going to launch Truffle and then use it's deploy function to start the init utilized by the smart contract. This makes it possible Truffle to identify mistakes that your code could contain at this point in time. As you work on your code, you're likely to want to take the time to run and run the compilation test using Truffle. It is to identify any problems prior to the final compilation. Truffle is also able to test the implementation of contracts within the virtual world.

Deploying

Once your contract has been tested and passed, you will be able to launch it using Truffle. To do this, start by logging into the console, and then using the command to start the truffle process (new directory) to create a new directory. Once you have this completed, you'll be able to find the contract you made by searching at the title of the contract as well as it's .solc extension. It is then time to locate an app.json/config file and insert your contract's name, where it reads "Contracts".

Then you will be able to begin the process of setting up your Ethereum node. Do this by opening a different window on the console, and then run to run the tesrpc commands. The last thing you have for is to run Truffle one more time, and then select the option to launch using your root folder.

Variables to be recollected

The variables you utilize to create your smart contracts will be organized in the same manner. The first variable you will encounter is the address variable that refers to the location of your wallet's ether, in the event that this is your primary wallet used by the contract. This address will automatically be created as part of the contract. You can locate it employing () function. () option. Utilizing Geth you can create multiple accounts in the same Node.

A second value will be UINIT which is a reference to an unsigned integer. It will always be referred to as 256. In the majority that of smart contracts you're not going to have a reason to change this variable.

The second variable will be listed as private or public, which allows you to select what your smart contracts will be able to gain access to. If your smart contract only has to use data that is being stored on the blockchain, you can change it to private. If you require the smart contract's information to be shared with be linked to information, you'll have to use what is called an oracle. You can then change the variable public.

Chapter 9: Ethereum And Smart Contracts

I'm going to assume that you've at least an understanding of the basics of cryptocurrency and blockchains, and how they function. If not, you should study before coming back to this. You must understand:

Keys for private and public keys

The reason why miners are needed by blockchains

Decentralized Consensus

Transactions

Concepts of smart contracts and scripting

It is also important to know more about Gas and the EVM.

The primary reason for Ethereum was to be an exchange technology for Smart Contracts, and they run using Ethereum's Ethereum Virtual Machine. This is a more efficient scripting language that Bitcoin does. Because these contracts run within EVM which limits the amount of resources used in each one, every operation is carried out by all nodes within the network. The contract code of an

Ethereum transaction triggers writes and reads of data, and sends signals to different contracts and may perform some costly computations, in addition to other things. Each operation is a cost which is calculated in gas. Each in the units of gas "eaten" through the transaction is paid in ETH. The price is determined by the cost of gas to ETH that is constantly and always changing. The price is deducted automatically from the account that is used to make the transaction. Each transaction also comes with an energy limit which is dependent on the amount of gas that the transaction uses. This helps protect against mistakes in programming that can cause your account to be depleted of Ethereum.

The Setting Up of the Environment

Once you have mastered the fundamentals now it's time to start programming. In order to begin the process of creating DApps You must be able to connect directly to the blockchain. This is your portal that lets you see on the Blockchain.

There are a variety of compatible clients , but the most well-known is the one called geth. However, it's not the most developer-friendly, so first, you should start using a testrpc.

To install and run testrpc start a command prompt, and type the following command:

"$ sudo npm installation install -g ethereumjs.testr

$ testrpc

As you begin to develop the testrpc program, make sure to open it in a another terminal and continue the work in another terminal. When testrpc is run it will generate 10 addresses. created, each one with test money being simulated to your benefit. You can do whatever you want using these addresses, and make use of them as a learning curve. It's not real money , so you won't loss everything.

Solidity is one of the most popular smart contract writing language , so it is the one we utilize, along with the Truffle the development framework. Truffle has been designed to assist you develop your smart contracts, create them, test them, and then deploy the smart contracts. Let's get started. Enter this into the command prompt (note you that lines starting with # are not comments but are intended to let you know the process we're using It is not necessary to input these lines:

The first step is to install truffle

The command sudo install NPM with a truffle

Now we'll create our project

Solidity-experiments with $mkdir

$ cd solidity-experiments/

$ truffle init

Truffle will then create the files required to build a project that is a sample which will include a token sample for MetaCoin. The contracts can be built using the command 'truffle compile on the command line. The next step is to transfer the contracts onto the network that was created using the testrpc network. For this, we employ "truffle migrate":

$ truffle compile

ConvertLib.sol... ConvertLib.sol...

The compilation of MetaCoin.sol...

Consolidating Migrations.sol...

Writing artifacts that can be used to ./build/contracts

The $ truffle is moving

Running migration: 1_initial_migration.js

Implementing migrations...

Migrations:

0x78102b69114dbb846200a6a55c2fce8b16f61a5d

Making sure that your network is successfully migrated...

Artifacts that can be saved...

Running migration: 2_deploy_contracts.js

ConvertLib deployment...

ConvertLib:

0xaa708272521f972b9ceced7e4b0dae92c77a49ad

ConvertLib to MetaCoin by Linking ConvertLib to MetaCoin

The deployment of MetaCoin...

MetaCoin:

0xdd14d0691ca607d9a38f303501c5b0cf6c843fa1

Making sure that your network is successfully migrated...

The artifacts are saved...

If you're using Mac OS X, you might encounter an error message regarding the .DS_Store file. Truffle gets confused by these errors, so you should erase your store files.

The contracts have been distributed on the network. Do you see how simple it is? Let's make an own contract.

A Smart Contract for the First Time

We'll be writing an Proof of Existence contract; the purpose for this project is to build an notary system that stores documents as evidence of their existence.

Create contract for $ Truffle to create ProofOfExistence1

Then, you can open that contract using Your text editor (use Vim, it's the most efficient) and paste the following code:

Pragma's solidity ^0.4.2;

/the Proof of Existence contract Version 1.

{contract ProofOfExistence1 contract ProofOfExistence1

/ state

bytes32, public proof

// calculate and save the document's proof

/ *transactional function*

```
function notarize(string document) {
proof = calculateProof(document);
|}
// Helper function to find the sha256 of a
document.
/*read-only function
function calculateProof(string document)
constant returns (bytes32) {
Return sha256(document);
|}
|}
```

We will begin with something that seems basic, but it is not correct. We will then go on to a more effective solution. This is an Solidity contract definition that is similar to the class used that is used in different programming. Every contract is a set of states and functions , and it is vital that we discern between two different types of functions that a contract can have:

Read-only functions are referred to as constant functions. They are those which do not make modifications to the status that the contracts are in. They only read, do any computations , and return results. These functions are solved local to each node and don't have any gas costs associated with them. They are utilized with an constant keywords.

Transactional - These perform modifications to the state of affairs and could allow funds to be moved. The changes have to be recorded within the Blockchain and this is done by the transaction to be sent to the Blockchain network. It also has an associated gas price. it.

The simple contract we're about to sign next can only hold only one evidence at any time using the data type bytes32. This is the exact size as sha256's hash. The function known as "notarize" lets you save documents hashes in a state variable known as "proof". It is a public variable and is the only way for a contract user to confirm that the document has been notarized.

The first thing we're going to use ProofOfExistence1 to deploy, but we will modify the migrations file to ensure you can ensure that Truffle can display the Contract. Replace the contents of

'migrations/2_deploy_contract.js' with this:
*
* migrations/2_deploy_contracts.js:
**
module.exports is function(deployer) {>
deployer.deploy(ConvertLib);
deployer.autolink();
deployer.deploy(MetaCoin);
// Add this line
deployer.deploy(ProofOfExistence1);
;|;| •}

You can also opt of deleting the information regarding MetaCoin and ConvertLib since we'll never ever needing them again. However, you should erase the test folder too in order to avoid failure. To initiate the migration, you can use these commands:

truffle migrate --reset

Smart Contract Interaction

We've built and implemented the contract. Now it's time for a call. It is possible to send messages to the contract using function calls , and you can examine the public status of the agreement. For this, we're going to make use of Truffle console:

```
$ Truffle console
// Retrieve the version that is deployed of
the contract.
truffle(default)>        var        poe        =
ProofOfExistence1.deployed()
Print the address
truffle(default)> console.log(poe.address)
0x3d3bce79cccc331e9e095e8985def1365
1a86004
// we now begin to register the first
"document"
truffle(default)> poe.notarize('A  brilliant
idea')
Promise { |}
// we now have the proof of the document
truffle(default)>        poe.calculateProof('A
fantastic idea').then(console.log)
Promise { |}
0xa3287ff8d1abde95498962c4e1dd2f50a9
f75bd8810bd591a64a387b93580ee7
// To check whether the contract's state
was correctly changed:
truffle(default)>
poe.proof().then(console.log)
```

0xa3287ff8d1abde95498962c4e1dd2f50a9
f75bd8810bd591a64a387b93580ee7

The hash is in line exactly with the one that we estimated prior to

Each function will result in a promise, and '.then(console.log) is used to print the outcome when the promise is resolved. The first step is find an image of the contract you have deployed and save it to the variable called 'poe.

The next step is when the transactional function known as "notarize" is invoked and it involves changing the state. The result is an unresolved promise which will be resolved to an individual transaction ID, but is not the return of the function. Remember that the gas has to be expended to alter the state of the system and transfer transaction data to the internet, and this is the reason we get the transaction ID as a promise. In our situation we don't care about the ID , so the promise is not considered. When you have written a professional application, you'll need to save it in order that every transaction is inspected for any errors.

Next step calling the function "calculateProof," which is, in fact, if you recall that it is a read-only function. Make sure you utilize the constant keyword in order to identify any read-only function in the event that Truffle is confused and will attempt to create a transaction to execute the function. The keyword is a signal to Truffle that we're not trying to communicate with the Blockchain and only reading from it, and with the use of the read-only option, we be able to get the sha256-sized document size, which is referred to as "a amazing idea".

This must be compared against the smart contract state therefore, in order to determine if the correct changes were implemented to the state we need to look at the public variable "proof". To determine what the public variables' value is we call a function which is named the same as it and this will result in an assertion. In this scenario your output will have exactly the same , and everything will work exactly just as it is supposed to.

Let's make an even better contract as the above contract is only registering one document at a moment.

Iterating Contract Code

It is the next stage to modify our smart contract to ensure that it supports multiple proofs. Copy your original file that is called contracts/ProofOfExistence2.sol and apply the following changes. The changes are that are changing the variable into one of the arrays bytes32 types and name it proofs. Then we make it private and include an application that will check to determine if the document has been notarized. This is done by repeating the array:

Pragma's solidity ^0.4.2;

/the Proof of Existence contract Version 2.

{contract ProofOfExistence2 contract ProofOfExistence2

/ state

bytes32private proofs

// Store the evidence of existence in the state of this contract

/ *transactional function*

Function storeProof(bytes32 proof) {function storeProof(bytes32 proof).
proofs.push(proof);
|}
Calculate and save the proof for the document.
/ *transactional function*
function notarize(string document) {
var proof = calculateProof(document);
storeProof(proof);
|}
//the function that aids in getting the sha256 value of a document.
*Read-only function
function calculateProof(string document) constant returns (bytes32) {
Return sha256(document);
|}
• Check to see to see if the document has been notarized.
* *read-only function*
Function checkDocument(string document) constant returns (bool) {checkDocument(string document) constant returns (bool)
var proof = calculateProof(document);

Return hasProof(proof);

|}

/return true only if proof has been stored

/*read-only function

Function hasProof(bytes32 Proof) always returns (bool) {*

for (var i = 0; i < proofs.length; i++) {

If (proofs[i =is proof) {If (proofs[i] == proof),

Return true;

|}

|}

Return false

|}

|}

Now we are going to interact with the functions - remember that you must update migrations/2_deploy_contracts.js so that it has the new contract in it and then run truffle migrate --reset

• Activate the contracts

truffle(default)> migrate --reset

• Download the updated version of this contract.

truffle(default)>   var   poe   = ProofOfExistence2.deployed()

Check for a new version of the document, you won't be able to find it.

```
truffle(default)>
poe.checkDocument('hello').then(console.l
og)
Promise { | }
False
```

Add the document to the proof store.

```
truffle(default)> poe.notarize('hello')
Promise { | }
```

Check to verify that the document has been notarized

```
truffle(default)>
poe.checkDocument('hello').then(console.l
og)
Promise { | }
True
// Success!
```

Other documents can be stored and recorded as well.

```
truffle(default)> poe.notarize('some other
document');
truffle(default)>
poe.checkDocument('some          other
document').then(console.log)
Promise { | }
```

True

This is a better version than the first, however it's not completely perfect. If we need to determine whether a document was notarized, it needs to be reviewed through all evidences that exist. A better method of storing the proofs is to use a map. Fortunately, Solidity offers support for maps, also known as mappings. Another thing we should modify is the removal of the comments making reference to read-only or transactional functions. The finalized version

Pragma's solidity ^0.4.2;

/Contract for Proof of Existence Version 3.

{contract ProofOfExistence3 Contract ProofOfExistence3

mapping (bytes32 = the bool) Proofs that are private;

// Store the proof of existence within the contract state

Function storeProof(bytes32 proof) {function storeProof(bytes32 proof).

the proofs [proofs] are true;

|}

Calculate and save the proof for the document.

```
function notarize(string document) {
var proof = calculateProof(document);
storeProof(proof);
|}
/the helper function which determines a
document's sha256
function calculateProof(string document)
constant returns (bytes32) {
Return sha256(document);
|}
```

• Check to see whether the document has been notarized

```
Function checkDocument(string
document) constant returns (bool)
{function checkDocument(string
document) constant returns (bool).
var proof = calculateProof(document);
Return hasProof(proof);
|}
/Returns true only if proof has been stored
Function hasProof(bytes32 Proof) function
hasProof(bytes32 proof) returns(bool)
{constant returns (bool)
Return proofs [proofs] of return;
```

|}
|}
It works in the same manner like the
second version we made earlier. Before
you test it, you should update your file,
and then run the 'truffle migrate-reset
command to do it again.

Chapter 10: The Present And Futures Uses Of Ethereum

Since Ethereum has grown in popularity and widely used, many have begun to explore possible ways it could be utilized in the future. As well, it's been adopted by numerous different applications which take advantage of the multi-faceted nature of this remarkable technology.

One of the most intriguing of them all are DAOs and DApps. They were initially described within the Ethereum whitepaper but have since evolved into a variety of types. In one instance there was an DAO related meltdown which led to the direction of Ethereum to shift forever.

DAO

What is what exactly is a DAO? DAO stands for decentralized autonomous organization. What is DAO?

A decentralized, autonomous business at its root is the notion that something, similar to an organization, is unsupervised and hardcoded. It is believed that a autonomous company is self-sustaining through the people who operate it, without any kind of leadership at the top of it. This could dramatically alter the way in which the business is managed.

A decentralized autonomous organisation is exactly what it's called. It is decentralized which means that it's distributed across several people, rather than controlled by one person or entity. It's independent, meaning that it is able to function independently and be significantly, if not completely automated. It's also an organization that can be explained.

The most adored DAO ever was the DAO. It was DAO was a disastrous project that ended up massively and transform Ethereum. What was the DAO?

The DAO was based on the concept of autonomous decentralized organizations and began using it. The method by which it was put into operation was to be made up of different smart contracts. These smart contracts were designed at the outset, and everyone who signed up to the service could benefit from an infrastructure already built.

What happens is that investors would invest into the service and get DAO tokens. The tokens could then be used by those who hold them to vote for various projects. The funds derived from the investment into DAO tokens could then be used to invest in the project.

This is where the concept of decentralized autonomous organisations is made apparent. It is believed that they let users be part of a jointly funded entity. In essence, the DAO was a huge crowdfunding campaign. But, it also served as a proof-of concept for a revolutionary idea in technology through the implementation of an infrastructure for autonomous, decentralized organizations.

Unfortunately, the DAO was to suffer a bleak fate. Although the infrastructure was amazing however, it had a fatal security vulnerability that allowed it to be invaded. In the last quarter of 2016, the DAO was breached.

To appreciate why this was such a huge deal, you must look be able to comprehend the dimensions and scope of The DAO. The DAO was, and is the largest bundle of smart contracts ever created. The project was staggering size and had a value of millions in investments in the period of the hacking.

But, over a lengthy time span, many Ethereum users were able to have their deposits in the DAO lost and the hacker was able to steal over a million dollars of Ether. The DAO rapidly fell apart because of this hacking attack, but it was not enough to recoup people's investments.

What did we learn? Well it appears that there was a split in the Ethereum community was pretty divided over the issue. The most obvious option is to just reverse the blockchain. But there's a crucial issue that needed to be addressed that blockchains weren't intended to function in this an approach. They're intended to be essentially immutable without any effort.

Although the enormous effort was certainly feasible however the idea that it was being contemplated was unsettling to a lot of Ethereum users. The notion of immutability, and not playing market prices is the main reason for the popularity of cryptocurrency. It was largely adopted from libertarians, who saw decentralized currency as a method of mimicking a completely inviolable market, which would not be distorted.

So when people began contemplating the idea of altering something that should be inaccessible There was a significant number of people who were against it. There were people who weren't opposed to it for motives of politics; others just didn't like the thought of having to duplicate the transactions that took place or in other ways lost money for example, in the time between the DAO breach and the rolling back.

The Ethereum founders were able to act immediately following the DAO hack and attempt to retrieve and restore the funds put into The DAO. But, there was many lost funds that were completely stolen during the attack. They were impossible to retrieve. The first option is called a soft fork. This is the process whereby an unintentional change in the blockchain happens with the purpose for rolling back the blockchain completely.

But, there is more that needs to be done in order to fix the issue that took place. The Ethereum community could decide to put the issue up for an election. The issue is whether it's more beneficial to restore the blockchain prior to the DAO's hack, and then recover funds, or if is it better to keep it in the same state as it was.

The folks who supported the latter were, in the majority the ones who believed that the incident should never be happening in the first place , and was nothing more than an incident that happened. They clearly would like to recover their money and were determined to make every effort to make sure that it didn't take place.

The supporters of the latter considered the blockchain to be indestructible. It also comprised largely of those who believed that the issue was the natural consequence of making investments in unstable technology. They thought it was not appropriate to alter things that have natural consequences.

In the end, the first won out. It was decided that the Ethereum blockchain was restored and the security flaw was repaired. The DAO would not last long following this event, but.

The same thing, in a way, resulted in one of the most significant events that have occurred in Ethereum past history: creation of a new alternative to Ethereum. The people who were off by the thought of altering the blockchain were extremely unhappy to see it changed. Because Ethereum is an open-source project it was decided to start a new version of Ethereum. This Ethereum fork will be referred to by the name of Ethereum Classic.

Ethereum Classic isn't the focus of this article obviously, but it's an essential and fundamental element of the Ethereum canon as it will have a large number of

Ethereum's users. Although Ethereum continues to grow in the years since the incident however, there are lots of people who choose to utilize Ethereum Classic, which is to all intents and purposes basically Ethereum but using the traditional blockchain. Although Ethereum is currently in second place when it comes to market value just second to Bitcoin, Ethereum Classic is in the eighth position. This is quite impressive when you think about how large of a slack in the cryptocurrency market actually is.

At the final point, it's easy for the tale of the DAO to make you skeptical of decentralized autonomous organisations however it isn't necessary. In reality, it was a negative application of a highly useful technology. The potential for decentralized autonomous organizations is beyond belief.

What went wrong with the DAO? The idea behind DAO is to be not centralized it's virtually impossible to alter the code after it's been deployed on Ethereum. Ethereum blockchain. That means the code will be running pretty much indefinitely. But, if there was to be a flaw within the program, there's practically no way to fix the problem.

What can we do to alleviate the situation? Nobody is quite sure. The best way to ensure security is to write highly secure code and avoid deploying anything that hasn't been tested thoroughly and with a high level of certainty that it won't fall as DAO did. DAO did. It's quite a lot easier said than done. However, this doesn't suggest that this is impossible.

DApp

DApp just stands for decentralized application. These are programs that are based in the Ethereum blockchain, which perform certain tasks in a specific way by running continuously.

In essence, they are simply smart contracts which are collected into decentralized apps which anyone can access. The lack of clarity in this definition makes an opportunity for something with lots of potential.

Strange and bizarre actions have been carried out using smart contracts! This is far more than the things you're thinking about right now. Let me give you a brief overview of some of the most intriguing ones. They all push the limits in the realm of possibilities Ethereum or smart contract are able to accomplish.

Consider, for instance, Etheria. Etheria is a recreation of the popular game Minecraft. It is an ethereal world that is divided into small tiles. It is possible to purchase the sections for just one Ether. But once you're in possession of a partition, no one else will be able to touch it. Additionally, you are able to make commands available to your tiny segment of Etheria by using clever commands. It allows you to construct and expand it with whatever you'd like. Although it's still new By visiting the Etheria website you can look around what's happening in the Etheria world and look at the tiny horses and the structures are being built by people.

It's also not possible to navigate this DApp section without opening EtherTweet.

EtherTweet is a large part based on of Twitter. It lets users send messages that are stored on and then retrieved from the blockchain. The messages are restricted up to 160 words. This is appealing to those who do not want censorship as the only ones who are able to delete a message in this platform are the ones who post the messages. This is one of the main reasons for the attraction of decentralization - because no person has the ultimate authority and, as such there is no way for anyone to interfere with the system.

One that is among the most controversial apps that could be appearing is TenX. TenX, should it roll out, will permit users to make use of their Ether coins anywhere can be used with Visa as well as MasterCard cards. It is possible to be used at any place which accepts Visa or MasterCard as they will be converted automatically to local fiat currency. This is unsettling for obvious reasons: it is believed that cryptocurrency can be mainstream and be used daily instead of being an obscure niche that is largely restricted by the internet.

Future Uses

As you will observe, Ethereum shows an infinitely increasing number of applications. It also has the greatest possibility of being adopted as a major cryptocurrency.

How does that translate to you as an investor? This indicates that although it might already be an expensive cryptocurrency it's likely increase in value. The best time to get into Ethereum is now. You just need to get started and then relax. What is the reason?

The Ethereum development team is among the most innovative development teams in the field of cryptocurrency and they also hold a significant amount in terms of market shares they are likely to take over the market.

If you're willing to go for it and invest in Ethereum go on into the following chapter.

Chapter 11: Security Risks To Blockchain And Ethereum To Prevent Any Mishap

Before you are able to comprehend the security features of Ethereum it is crucial to know what kinds of threats that exist on the internet.

The types of attacks that are available can be classified into passive attack that observe communications as well as active and close-in attacks designed to take advantage of the information contained within an information system. Certain attacks are carried out by service providers, and the information system can be a tempting to hackers.

Passive Attack

It's an attempt to track unencrypted internet traffic and to search for sensitive data and passwords in plain-text. The information is used to other attacks. The attacker can analyze the communication, look for unprotected messages as well as decrypt encrypted communications and obtain passwords. Passive intercept can be beneficial in helping adversaries to see the coming actions as well as other actions. This can expose files of data to other attackers, without the knowledge of the victim.

Active Attack

In this kind attacks, an criminal could evade security by using the use of stealth, viruses Trojan horses and worms. These kinds of attacks aim to break security functions and malicious code are created to steal or alter the data. Active attacks are targeted at the core of the network and also the information flowing through the. In such attacks, attack on an enclave or attacks on remote users who is connected to an enclave is carried out. Due to active attacks, information files become exposed DoS as well as the data gets altered.

Distributed Attacks

In this attack, code introduced by the attacker are employed, including Trojan horses or a trusted part of software to be distributed to different customers and businesses. In distributed attacks, malicious changes are made to the software and hardware in the factory or in the process of distribution. The aim for these hacks is to introduce malicious code as an access point for devices to gain access to them without authorization in order to obtain information later.

Close-in Attacks

It's a method to physically connect to the various components of a system, network and systems to gather the most information possible about the system. This could involve regular users who live near network and system. The attackers attempt to steal private information, and alter any password or data in order to block access for authorized users to data. The network is accessed through a covert way. can be beneficial to gain close to the target and gain access.

Social engineering is an infamous type of close-in attack since the attacker gains access to the data of a specific person via interactions with them. It could send an email or make a phone call. There are a variety of techniques employed to divulge information regarding security at the workplace.

Insider Attack

In this kind of attack, someone gets at risk from the outside for example, a disgruntled employee. The attack may be malign and not, depending on the motives of the attackers. Insiders who are malicious try to spy or steal data. They can then use it to defraud or limit access for the authorized users. Inattention, ignorance and deliberate evasion of security aren't the reason for this kind of activity.

Phishing Attack

In these kinds of attacks, hackers create fake websites that look exactly like the websites such as PayPal or banks. The hacker then sends out an email message that entices users into clicking the link which takes them to the fake. If the user attempts to login to their account, the details of their account is stored by the hacker, and the user can verify to access this information via the authentic website.

Hijack Attack

In this kind of attack it is when the hacker hijacks an interaction with your customer, and stops communications with other people. The hacker speaks to you the same as the other person, and you believe you're communicating with the other party. You can also send confidential messages directly to hackers.

Buffer Overflow

It is a type of attack where an attacker provides an application that is beyond the capabilities of the application. The attack could lead to gaining admin rights to the application via an command prompt.

Password Attack

In this kind security breach, an criminal attempts to hack passwords in order to gain the access you need to personal data. The password database is their primary target, and they are attempting to steal passwords to access secure data. There are three main kinds of password-related attacks like:

* Attack on the dictionary
* Hybrid attack

* Brute-force attack

For a dictonary-based attack it is a list of words which could constitute a listing of possible passwords. A brute-force attack is a method to join characters.

Exploit Attack

Exploit attack is a kind of attack where the attacker has complete knowledge of security issues of operating systems and the software. The attacker can exploit weak points to gain expertise.

Spoof Attack

In these kinds of attacks, hackers alter the source address of packets that someone sends. The receiver interprets them as they came from a different source. It's an attempt to circumvent your firewall's rules.

Routers, firewalls, as well as switches are essential elements in the networking. They are the gatekeepers of networks to protect your servers and your applications from intrusions by hackers and other intruders. The most significant threats to the network include:

* Information gathering

* Sniffing
* Deny of Service
* Spoofing
* Session Hijacking

Host Threats and Countermeasures

Host threats pose danger to the software and applications running on your system. For instance, the Server that is part of Microsoft Windows 2003, Internet Information Services (IIS), Windows 2000 and.NET Framework are crucial to safeguard against threats to host systems. Host threats at the top of the list are listed below:

* Trojan horses, viruses, and worms
* Profileing
* Password cracking or creaking
* Refutation of the service
* Random or arbitrary execution of code
* Access to the Internet without authorization
* Footprinting

Viral infections, Worms, and Trojan Horses

The worm, virus as well as Trojan horses, are harmful software that could cause disruption to your operating system as well as applications. They can also attack operating systems and the applications.

The worms, viruses and Trojan horses may attack via inadequate defaults or software bugs, mistakes of users, and weaknesses in internet protocols.

Countermeasures

Keep up-to-date with the new OS and Software.

* Block unneeded access to all ports through the firewall and the host.

* It is crucial to shut off idle functionality for example, protocols and services.

* Modify the default and weak configuration settings to ensure it is safe and free of all dangers.

Food Printing

It is a well-known method to obtain information from a computer system via DNS queries such as ping sweeps, port scanning, and World Wide Web spidering.

Hackers can gain access to all of your personal details by starting with basics, such as names and email addresses, among others. They are able to monitor your IP address on your site and look up your servers online.

Countermeasures

It is possible to avoid this by limiting the responses requests. Be sure to turn off unnecessary protocols, and then evaluate the data prior to publishing it on the site.

If you are looking to make your system safe from security threats, it's crucial to eliminate all unneeded programs, deactivate other protocols, ensure that you use strong passwords and check the URL of websites prior to entering sensitive information, and strengthens your TCP as well as the IP stack to prevent the possibility of denial of service. It is important to set up IIS to stop URLs "../" as it will assist you in avoiding any arbitrary code execution. Always set secure web permissions and secure folders and files with the assistance in the case of limited NTFS permissions. Utilize the .NET Framework access control mechanism.

BruteForce Attack

A password is able to be cracked using the Brute-force attack It is a method that makes it is possible to combine numbers as well as special characters and letters are utilized. The time of attack may depend on the level of complexity of the code.

Countermeasure

If you wish to stay clear of this kind of attack, be sure to avoid complex and long passwords and also try using upper and lowercase letters and numbers. It can require thousands of years to crack a complicated and lengthy password. For example an example, a password such as "ieatfood" is simple to crack, whereas an account like "aP85KL31" is hard to get past.

Social Engineering

Social engineering lets you manipulate someone to believe you, so that you are able to obtain information from them.

Example 14

An attacker may try to obtain an account password for a colleague or friend by calling the victm as an expert from IT department. They will ask for login information. The hackers can contact the victim, impersonating an official of the bank, and ask for the details of a credit card. It's used to gain an account password, bank details as well as personal details.

Countermeasures

If someone called you as a bank representative and asked for your information about your bank or personal details you should ask them some questions. Be sure to verify the authenticity of a person, and don't provide any information about credit cards personal information, private data or cell number on a phone call.

Rats and Keyloggers

In keylogging , or rating, hackers send keysloggers, or rats, to the victims. The keyloggers allow the hacker to observe every aspect of the victim's computer. Every step is recorded for security reasons, including bank information and passwords. Hackers can take control of the computer of the victim.

Countermeasures

If you wish to stay clear of keyslogging or rats it is not necessary to sign in to your bank account through a cyber cafe or on the computer of someone else. If you are required to utilize a virtual keyboard make sure you use an anti-virus program and update it regularly.

Phishing

It's one of the most popular and easy methods that hackers use to steal the account information of an individual. In these hacks, hackers create fake websites of the genuine websites like Facebook, Gmail, etc. When users log in to the fake site hackers can take the login details and put them the genuine websites. It's easy to create fake pages using the aid of free web hosting websites.

Countermeasures

It is easy to stay away from phishing attacks by examining the URL as the URL of websites that are phishing differs from the original URL.

Example 15: The URL of the fake facebook could be similar to faccbook.com. It could be a different character , as you will see "c" in place of "e" within the URL. Be sure to verify the URL to see if it's correct, fill in your personal information.

Rainbow Table

Rainbow Table is pre-constructed list of hashes, as well as any other combinations of characters. The password hash is processed using an algorithm that is mathematical like md5 could be changed into something completely new which isn't recognizable. It is a method to encode the password. After it has been encrypted there is no way to retrieve the original string. MD5 is the most commonly employed hashing algorithm that stores your password on the site. It's very similar to dictionary attack with just one difference : the rain tables can only be used to hash characters employed in passwords. In contrast the dictionary attack is based on are the normal characters that are that are utilized in the password.

"iloveu" in md5 is edbd0effac3fcc98e725920a512881e0 and

for hello, it is 5d41402abc4b2a76b9719d911017c592.

Countermeasures

Be sure to choose a lengthy and complex passwords because it can take a lot of time and effort to make a table to store long and complicated passwords.

Making guesses

It's a flims technique, but you could apply it as it could help you obtain the password within less than a minute.

It could be carried out by hackers who know you and has information on your password pattern. He could guess the password or employ social engineering to obtain the password.

Countermeasures

There is no reason to make use of your own name, number or surname, birth date, birth or roll number to create passwords. You can make your own password that doesn't be associated with your personal information. You must create an elaborate and lengthy password that combines alphabets, numbers.

Cookies and Identifiable Theft and Countermeasures

Information that is personal to you can be stolen from your system through the aid of cookies. If you visit the internet and a cookie is created, the website is automatically saved to your computer. The hackers could steal this cookie to gain access to your computer easily and take sensitive information.

If you browse a site and then a cookie is created, it will be saved to your computer so that your data can be easily recovered from the website upon your next visit.

Countermeasure

If you're looking to secure your system from being hacked by a cookie and identity theft it is crucial to keep your PC and mobile devices clean. Install a reliable antivirus software like the AVG or AVG and Avast and perform a thorough system scan. Keep your antivirus up-to-date to eliminate regular security threats.

Stride threats and counter Measures

Threats to STRIDE are fought by applications since hackers hack your computer to gain access to important data. The term STRIDE is a reference to:

Spoofing

It's an attempt to gain access to the system with the assistance of a fake identity. The hacker typically steals credentials or uses an unauthentic IP address. Once they have gained login as legitimate hackers abuse their rights and gains access to personal information.

Countermeasures

Always make sure to use secure authentication, and do not store passwords that are secret in plaintext. There's no reason to transmit your passwords over wires. You can secure authentication cookies through using secure socket layer (SSL).

Tampering

Tampering is a shady manipulation of information. like data flowers in a network that connects two computers.

Unprotected data packets could be damaged or interrupted. In the event of malicious code execution, it is possible to cause data corruption.

Countermeasures

This can be avoided by using data hashing and signing. Digital signatures are a good option to use and increase the security of your authorization. You can make use of security protocols that resist tampering across communication channels.

Repudiation

It's a capability of the user to deflect the execution of a particular operation. It is impossible to prove the existence of these attacks without an audit.

There's a web application that allows security and control of access based on the concept on "JSESSIONID". The actions of an authenticated user is determined by the parameters of the users and determined by the cookie header.

Countermeasures

To stop this type of attack to prevent this attack, you can establish secure audit trails, and also use digital signatures.

Information Transparency

This is a wrongful disclosure of personal data . It could be caused by poor handling of information or insufficient security measures.

A person can read the content of a file or table, or cannot open or examine the data information is provided in plain text over the network. This could be the result of the use of hidden files or string-like comments embedded inside a the database connection, or even poor processing of the data. This can lead to increased external and internal threats. Additionally, the information you have stored could be disclosed to your adversaries. The information could be beneficial to protect themselves from hackers.

Denial of Service

It's a procedure to render a system or application inaccessible.

Denial of service can be created by a bombardment of requests to servers making utilization of all resources of the system. Input data may be altered to disrupt the operation of the application.

Countermeasures

Techniques for throttling bandwidth and resource usage are a good way to stop attacks that cause denial of service. You can apply filters and validate inputs to solve this issue.

The Elevation of Privilege

A privilege that is elevated means that users with restricted access could assume the role of a privileged user gain access to an application, and thus enjoy maximum benefit.

An attacker who has limited access can increase the privilege level for conciliation , and also gain control over a highly privileged and trusted accounts, or a process.

Countermeasures

Follow the guidelines of least dispensation and utilize accounts that are least dispensation in order for the execution of the procedure and access the available resources.

Chapter 12: Could Ethereum Be The New Bitcoin?

The reasons why cryptocurrency is so fascinating is a difficult thing to explain in a couple of phrases, but the reality is that cryptocurrency will play an important part within our daily lives into the near future. The question is when we can pay for services using virtual currency (crypts). When we all make transactions on stock exchanges using cryptocurrency, and when it becomes equal to money. Don't fall for the illusion that it will be the case sooner or sooner. As of now, Bitcoin as well as Ethereum try to establish themselves as currency in a number of markets, and it will take place. If this occurs, you will see an influx of other cryptos that will be comparable to currencies and will be internationally acknowledged.

We believe this could take place in Croatia however, the law currently does not recognize Bitcoin or other currencies as a currency or value. It isn't even a tax for trading in bitcoin that is an frequently asked question for Tax Administration. Tax Administration. The recognition of cryptos will begin at the level of Japan, China, and America and then later in Europe and eventually in Croatia. This article won't focus on it. It will instead be discussing instead one of the hundreds of cryptos in the market is Ethereum is a crypto with amazing plans and greater potential. Many people believe that Bitcoin as the one legitimate currency that is worth talking about however, it's far from the reality. Etherum is crowned with impressive strides. The issue is when will he be able to surpass Bitcoin.

It is important to differentiate Ethereum in this regard, namely, Ether as an encrypted (virtual coin) and the Etherum platform, on which Ether is based. Ethereum is a publicly-owned open source, blockchain platform that uses smart contracts, also known as smart contracts. However, before we dive into more specifics, we'll try to describe in simple terms the way Ethereum operates. When you access any of these services (for instance, a web service, social network ...) to save personal data, it is stored on computers, or servers that are owned through Google, Facebook, Amazon or any other neighborhood. The companies that run Touch have a large number of staff specialists who are concerned about the security of these servers, their servers as well as data replication and the like. However, this is a server-client model that is extremely vulnerable.

If the server is down, or it's not working then nothing else can work. You are unable to join Dropbox and access your files. You cannot access Facebook and send an email to a friend or access emails via Gmail ... Every firm has just one fault or important area.

The creator of Apache Web Server said - it was the first step and the very first sin committed by the internet. This model is becoming increasingly difficult to maintain today, with all the attacks and hackers that are executed. There must be some form of decentralization for servers and data. Ethereum is on the right track.

In this system, currency known as "Ether" are created to allow nodes to compensate for their resources. If you're a member within the Ethereum network, you are rewarded for each transcription entering blockchain data a certain amount of Ether. This is the ether of the crypto that climbed dangerously upwards, and whose has a huge increase in value and can "download" Bitcoin from the top of the most sought-after cryptographic crypts and surpass its value. The Ether value is $ 280, though the value of it varies. In spite of the fact that the value last year was $0.20 0.20 There is an enormous increase. Additionally, investors who made smart investments on the stock exchanges crypts have a decent amount of money today. It all depends on the amount they invested in.

If you put $100 into 500 ETH in the 8th of last year. Today,, $ 100 is about $140,000. Yes, you read correctly. The same is true for Bitcoin. Anyone who made a bet on

BTC 5 to 6 years ago could be able to make millions due to 1 BTC equals 2800+ dollars. However, we didn't take advantage of the chance for the majority of us. However, if you decide today , to invest money into Ethereum and there is a high chance to make a 10 percent profit over the next two. According to all the research, by the end of the year, Ethereum should arrive and surpass Bitcoin this means that the worth of Ethereum could reach 3000dollars or more. Let's not forget that every node in the network is granted advantage of certain guidelines (smart contracts) that are based on the principles of. If a specific event triggers a smart deal that is then shut off. If everything is in order and the default is set it is, the node of blockchain (distributed database) writes data. That's what every node is performing.

You have a decentralized network in which every node is aware of everything and has an entire database that holds all the information. The only way anyone could "tear away" this network would be to gain control of more than 50% from the Ethereum network in the first place, which is practically impossible. In the end, an intriguing idea that is bound to change the web we have come to know. This isn't an opportunity to invest money you don't have, but if have money you could lose, investing in Ethereum may make sense. If you invest $1000 it is likely that you will have $10,000 in a short time. The risk lies with you, as Etherum could fall to zero, and all the money you've that you have invested in it is ineffective. Therefore, invest wisely.

The most important question is what happens to Ethereum is going to go? The platform is only two years old, however over the past two years, they've done many things. However, they will be doing more in the near future. As you can see from the document that they are currently working on cryptography, security improved and more flexible smart contracts, proof-of-work mining innovative applications, the development of new algorithms and more. Each new feature will mean that the value of Ether will increase, however Ethereum will also expand in such a way that a variety of applications will depend upon it.

A lot of countries are thinking about using the Etherum platform to implement national projects, such as accounting as well as land-ownership. Why shouldn't it be recorded as a database distributed across the globe, and what makes it not clear? Additionally, there are logistics initiatives that can cut down on costs , which range from millions to 10 million dollars. There are also smaller projects on which we only have details. We also recall Uber. Why shouldn't driving , and all money transactions be stored in the form of a blockchain, with every driver was a knot that would record it? In reality, the sky is the limit of this kind of technology. Whatever you imagine is possible similar to that. It is also possible to look into Sisco's project. It's an encrypted and distributed Dropbox which will be decentralized , and may prove to be exciting in the near future.

Interest in Etherum was sparked just a few days ago in Russia. The chief programmer and chief of the Etherum Are Vitalik Buterin, met with one of the most powerful persons in the world , Vladimir Putin. While there's not any information on what they discussed however it is clear that Putin was keen on the possibility of using Ethereum into Russia and was in favor of certain initiatives. The oil and gas industry brings huge amounts of money into Russia however, they also recognized the need to expand the scope of operations and that the digital economy is can be the base for new tasks to be addressed. It's also interesting that in Russia the pilot project for the Etherum bank is now complete and Sberbank is also a participant in the project. This suggests that Russia may be able to move ahead in America and be the first country of its size to accept the cryptographic payments as money, and to use it to conduct faster and more secure businesses, such as making and receiving payments as well as conducting business with corporate entities. In the end, there won't be any fraud, non-payment or other issues, or, at the very least, less of these. In any event the news about Ethereum is being discussed and it will be fascinating to watch. Particularly during in the next year, there is a massive growth of Ethereum as well as other currencies.

Advangates

Integrity (unchangeable) The third party cannot make any modifications to the information

Consensus – applications are based on the principles of consensus. It is not possible to censor and the decisions are taken in a unanimous manner

Security - Without a central hub, that can be the bottleneck, and when it's not working and the entire application goes unusable. The applications are extremely secure and cannot be compromised.

They're always online and will never cease to be unavailable or shut off It is impossible for anyone to stop it from running, even the government can't stop it.

Naturally that not all of them are perfect and the numerous advantages of decentralized apps are not flawless. The smart contract is source code created by developers, and they're exactly as good as the great and experienced programmer who created them. Incorrect code that are made by the developer may result in actions being carried out however they weren't planned for and therefore not anticipated. If there's a glitch in an application which is later used to harm it is not possible to find a solution to correct the mistake. The only option is in accordance with principles of consensus everyone involved in the network will agree and the source code gets changed. This is in direct contradiction to the integrity that is guaranteed by the blockchain and any action that is in violation of the fundamental rules of decentralization.

The price of Ethereum has seen an enormous increase over the last year, everybody has been talking about Bitcoin however Ethereum is a step ahead. In the past it was possible to purchase Ethereum for just $10, however, now you have to put in more than $1000 to buy this. Even though Bitcoin is now becoming more costly for many people, Ethereum has the potential to reach $ 2,000 and is still inexpensive. Decentralized apps are created from the source code which runs and runs on a blockchain-based network. it isn't feasible to regulate its behavior against a centralised organization.

Every centralized service or application could be decentralized by Ethereum. Banks can withdraw credit and registration of users in any kind of register, voting online.

If you are looking to purchase Bitcoin Cryptocurrency from Serbia the shortest description of the scenario is:

* Option of exchange for the crypts

* Funds deposit (money in dollars or euros) via PayPal or credit card

* Exchange of money for Ethereum

* Switching crypts over to the safer area (wallet)

The description was brief on the product. We are now moving to provide a thorough description of all of the steps of shopping and the things you should pay attentionto!

There are many of exchange centers where you can make the exchange and purchase of the crypts. All exchange offices are classified into:

Cash Exchange - This group includes ATMs and retailers.

They have installed ATMs that you can purchase and sell crypto currencies. The issue with the Mana ATM is that there are only a handful of ATMs in operation and often Bitcoin is not to purchase.

However there are websites where you can contact retailers from Serbia or the city you reside in, that can sell you to cryptocurrency in exchange for cash.

Internet Exchange Offices - This group of websites includes sites online where you can buy encrypted coins by using either a credit card or Paypal.

Here are a few aspects you must be aware of when choosing the currency exchange you want to purchase and buying these:

1. Privacy. If you are looking for privacy or are not willing to share personal information when you shop and online, then skip online websites and instead shop with a credit cebit card, or Paypal. In particular, these websites require verification of your identity prior to purchasing that involves providing them with overview of all of your personal details, including scans of your personal documents. This is a standard procedure that isn't essential to be worried about this since they all work in this manner However, it might be difficult to comply with this requirement since the process can be slow and therefore it is important to remember this. In this scenario the best choice is cash exchange, which is, buying from ATMs or directly through traders.

2. Limitations. This is only applicable to people who are looking to purchase an amount cf crypts in one go. In particular, all exchanges have specific limits that permit orly certain amounts of crypts that can be purchased. As we do not have the

same payment capability from Serbia This will not pose any issues for purchasing.

3.Speed.The purchasing process isn't always as simple. For instance, as I've previously written that online exchanges require verification of your identity of the buyer before purchasing in order to be a lengthy process for customers who wish to make purchases previously and this could take a few days. In contrast buying at retailers or ATMs is the fastest way to go when funds are transferred, crypts are delivered to your bank account.

4. Exchange course. There isn't an official and fixed course for the crypts. It's based on demand and the price based on the exchange office that is an exchange bureau. The most useful website on which you can keep track of the value average of all cryptographic crypts is CoinMarketCap. You can also determine for each crypt which exchange you should make purchases (where is the most effective route).

5.Scams.You must be extremely prudent when purchasing. As with all places there are plenty of scam alternatives. This is why it is essential to select the most reliable exchange companies and find by the number of transactions processed by an exchange company to date. But, I should be clear that it's essential to not store your crypts in longer routes in exchange offices. It is recommended to transfer them into your account immediately after buying.

6. Commissions. Every exchange has its own commissions and it's important to know what the commission you will be charged prior to making a purchase as it could be quite excessive. The commission is usually paid for each step of the purchase. This is when you pay for your money in exchange for crypts, as in the transfer of crypts into your account.

The crypto wallets are equipped with secure keys, or private codes. These permit you to alter your wallet and the money within your wallet. Every wallet creates a public address, which is linked to a particular cryptocurrency. It is a unique address, and you'll be using this address when you wish to get steam from someone else (for instance, if you buy crypts through a currency exchange, the exchange will transfer you the currency using the address). To access your account, you'll create a security code which you'll only have. It can be saved on your computer, as an image file, or print it out on paper.

There are many types of digital wallets.

Physical (Hardware) wallets are tiny devices that are similar to USB sticks. They are designed to function as your wallet for encryption. In order to work, they require an internet connection typically through an USB port. Most popular include: Ledger Nano S, KeepKey and Treasury. I'll go into more detail about the three in a separate post. The most important thing to consider with this kind or wallet is they're an excellent option when it comes to security. They are very easy to use. There is only one drawback: you must purchase the coins. It is advised to store more crypts with higher values in an account. Security codes are stored inside the wallet, therefore only the owner has access to their wallet, if the owner has a password that allows him to gain access to your security number.

Internet banknotes - these are applications that can be used to create banknotes on phones, computers or tablets. They are extremely simple to use and highly recommended if you are looking to store fewer crypto currencies inside your bank account. This security number is saved on the device on which the wallet functions, therefore the wallet isn't 100 100% safe.

If you have invested a significant sum of money on purchasing crypts and wish to hold them in the long term, digital wallets might be the ideal option for you. If, however, you wish to trade crypts then an online wallet is the best option to meet your requirements.

In this article, I would like to point out that you should distinguish between the internet wallet and the online exchange. In both, it's likely that you already have your own cryptographic crypts however they are safer when it comes to your banknotes. Exchange offices can be risky since nobody can assure you that they won't be able to burn them out due to any reason. Therefore, your crypts may disappear.

Another point. There are over 1000 cryptocurrency wallets available, alongside Bitcoin and Ethereum they are most sought-after. It is crucial to note that every wallet doesn't allow the storage of every encrypted crypt. So, based on the wallet you use you'll be able to save just one or two distinct crypts within one.

Chapter 13: Ethereum Use Cases

It is important to note that the Ethereum technology is in its infancy. This means that any company and tech enthusiasts who are interested in
being aware of the effects of technology is essential. be aware of the latest developments in technology as well as instances of its application in industries.
Here are a few of Ethereum's usage examples:
* Smart Contracts
* DAOs
* Dapps
Let's get started and look at these use cases.
Bright Contracts3right contracts execute themselves using algorithmic codes which are used to store and reproduce on dispersed ledgers
system, the Blockchain. The algorithm is run through the network of nodes that are running the Blockchain and could cause regular

changes to the ledger. For a different perspective smart contracts are an application that runs on the Blockchain when an activity occurs.

triggered.

To allow the application to function it needs to be verified by a variety of nodes of the distributed community to confirm the authenticity of the application.

If you're thinking about using the Blockchain technology as well as its power to distributed storage that is reliable smart contracts

can provide secure computations using the storage dispersed.

With smart contracts, there is no need for a single source of management. Bright contracts utilize blockchain technology. Blockchain

technology in which multiple parties, autonomous computers, use a an algorithm that continuously checks and verify updates

into the ledger. This increases the transparency of the ledger.

Since all the nodes on the Blockchain network are running exactly the same code each one of them is verifying the other smart contracts

The information will be available to everyone. Every node is able to look at a smart contract and if it is satisfied with the reasoning the contract, it is able to utilize it. It is possible to use it.

On the other hand, when the node isn't in agreement with all the code the node won't execute it. This is how the concept of transparency gets promoted.

contracts that are intelligent.

Smart contracts may be advantageous for a broad variety of companies, including banks and healthcare providers, as well as insurance

companies. If properly managed businesses, they can gain from fewer risks in real-time processing, accurate and validated

transactions, less third-party involvement and lower costs.

Smart contract can be developed by using Solidity, the Solidity Language as well as

Pyethereum programming languages. Smart contracts are programmed

They only execute when they receive instructions, which could take the shape of a transaction within the EOA. The contracts could push or

Pull funds and request the necessary actions from various contracts, while calling the contract code to carry out tasks that are

dynamic. Here's a good example of a smart-contract code:

Pragma solidity ^0.4.0;

Contract MySimpleStorage

storedData;

Function place (uint mydata)

MyData = StoredData

function Get () continuous returns (uint)

Return storedData

DAOsWhile smart contracts by their own are intriguing It is the idea of an array of uni-directional contracts that are operating

Collectively, it showcases the vastness and potential value in Ethereum's tech. In conjunction with DAOs (Distributed

Autonomous organisation) (also known as DACs (Distributed autonomous corporation) Smart contracts may help to enforce the principles of
the ecosystem.
Here's how DAOs function:
* A set of nodes write"smart contracts" (codes) which will manage the business or organization.
* There's the initial financing period, in which nodes invest funds into the DAO by purchasing the parts that are ownership (this is
called an audience sale, or Initial Coin Offering (ICO)) to provide it with the money it needs.
* Once the funding period is over after which the DAO begins to operate.
* Nodes will then be able to make recommendations to DAO regarding the best way to use the funds, as well as the nodes that have been bought by can take part in the vote.
to accept the suggestions.

As of now all Intelligent contracts we have discussed are owned and operated by different reports, which we believe to be

were humans. There isn't any prejudice against robots, or any other individuals within Ethereum. Ethereum ecosystem. Particularly, the wise.

Contracts can trigger random actions in the same way that any other account.

They may own tokens, take part in audience earnings, and be voting members of various contracts. DAOs are able to

Facilitate such arrangements within these arrangements within the Ethereum blockchain. The way in which this particular democracy functions is that any intelligent contract code

There must be an owner who acts as an administrator.

The Owner may be able to add (or remove) members who vote in the group. Any member can make proposals to the ecosystem and it is a matter of

in the kind of contract trade that allows you to either ship the Ether or to execute a smart contracts. The members then have the option of voting for or against the trade.

Refuse the idea. If a predetermined time frame is set and all members have been able to vote, the contract is counted the amount

votes , and if they are enough votes, the company is going to go ahead with the transaction.

DappsThe main objective for the Ethereum system is to function as a platform to facilitate the development of distributed applications

(Dapps). Dapps can be created from an individual DAO or a sequence of DAOs working in concert to develop an application. This could lead to

Similar to apps like Google Chrome or Microsoft Outlook.

These apps are designed to meet a particular function. But, in order for an app to be classified as an Dapp that it meets the requirements of

Following criteria:

*The app must be completely open source and operate independently of an entity that manages all of its tokens. The app can be adapted
the protocol to respond to the improvements planned and to encourage feedback. However, any changes must be approved by a the majority of its members.
of its of its.
* The data and records of performance should be cryptographically kept in the hands of individuals. Blockchain must also be distributed in order to stop
the most fundamental reasons for failing.
It must use cryptographic tokens to run the program. Any worthless contribution from miners must be rewarded in
app tokens.
It must create tokens based on a standard algorithm for cryptography that serves as proof of their value.
Dapps are classified within four types:
Smart contract utilities, smart contracts and analytics

Information validation, and Oracle solutions

* Gaming and gambling

* Corporate governance and registry.

Chapter 14: Ethereum - Decentralized Application (Dapp)

The work of many developers is centered about learning languages, new platforms or even the frameworks. The most interesting part is when a developer is introduced to an entirely new approach. One of the most modern and most technologically advanced paradigms are blockchain's decentralized networks.

Because it's a completely new paradigm, we're going to explore the various technologies needed for the consensus network as well as analyze what contributes to the network possible.

Main technologies

* Hash Function * Hash Function Cryptographic

Hash functions can take a chunk of data and convert it to data that is of a certain size. For instance an image of a 2MB file is sent to a hash function will produce two hashes that are 128 bits in length. An Cryptographic Hash function executes the functions and fulfills three important specifications:

The information provided is not specific in relation to the non-reversible hash generated from the data input.

minor changes in the changes of input create an output hash that differs in a manner that the hash cannot be determined using the function of hash.

A very low probability that two inputs could make the identical hash.

Key Cryptography for Public Key Cryptography

It is an encryption that requires two distinct keys The "private secret key" is reserved intended for the person who owns it, while"public keys" are for anyone "public key" can be used by anybody. There are a variety of features that are beneficial in key cryptography.

1. Encryption of information is performed by anyone using the public key and an encrypted private key in order to unlock it.

2.The the private key owner's capacity to sign with private keys and have the data being verified by the person who holds an open key, with no the disclosure of the information. This is usually used in the systems of a DCN's account. It's used to establish the basic principles of transmitting transactions.

P2P Networking

The computers connect directly to each other without the need to send requests to servers. The computers on the network are known as peers, and are all of the same status in the same way as each peers. This network is built on the an altruistic spirit among the peer group and they share all the resources in the network.

Blockchain Technology and Crypto Economy

The Blockchain

It's similar to a database type, which is utilized in the DCN. Information can be stored and rules are established based on the latest information. It is generally updated by chaining blocks together using hashes derived from the block's content from the prior block. The block contains the most actual and historical data.

Transactions or requests are used to modify the database's state The blockchain also stores the signature.

Documentation of the work

Evidence of work was once an anti-deterrent mechanism. It's an easy way to prove that an enormous operation has been completed. It is often executed through ish function (cryptographic). If you're provided with certain data elements, (e.g., block header or the list of transactions) then you must find the second part of the data and then, after combining to the primary, a hash which is unique (e.g. certain zeros that trail) is generated.

Technologies within Ethereum

EVM (Ethereum Virtual Machine)

This is a significant technology that is designed to be utilized by anyone in a network. The EVM can read/write executable data and codes and assist in the verification process of digital signatures as well as execute code in a way that is almost like a Turing. When a message has been verified by a digital signature and the blockchain information confirms that it is safe to use it, the code is executed.

The Generalized Blockchain as well as the Decentralized Consensus Network

As we mentioned earlier, Ethereum is a peer-to-peer network where each peer keeps copies of the database. They also run an EVM that manages and changes the state for the Blockchain database. The incorporation with Proof of Work in the blockchain technology allows new blocks require participants of the network to participate in Proof of Work. Incentivizing the network provides the consensus needed for members to agree to the longest blockchain through using the distribution system of "ether' that is an cryptographic token.

Following this, we're left with a method which is not suitable for the p2p network or the client-server network since its state isn't consistent. Due to the cryptographic security and its distributed nature it could be an outside party which isn't trusted and does not interfere with outside entities.

The decisions made by cryptocurrency have had a financial consequences for organizations as well as individuals, and also other types of software.

The developers have come up with different methods of allowing different components of software and the internet connect. We'll discuss the advantages of decentralized app.

Development environment set-up

Web designers can have a simple design with Ethereum since the language used in development is well-known to everyone who has JavaScript. There are three parts of software that developers download

AlethZero

It's an GUI client with advanced features that include forces mining, private chain as well as a full WebKit.

Mist

It is the DApp browser as well as the mining client that is used by the client for access to DApps.

Mix

It's an integrated environment of development which builds and compiles contracts in conjunction with their front-ends.

Software Requirements

There are three software components that we've discussed previously that require download.

The first thing to download the latest AlethZero stable version, a C++ client, and any operating system. If you encounter issues with the build, you should upgrade to the latest version that may have solved the issue.

The second step requires you to install MIX an integrated development environment available on both Mac or Windows. You may also install MIX on Linux.

The final step is installing Mist in DApps and then tweak the front-ends whilst creating.

Extras

A Mix or text editor will make the contract code on the backend, which we'll be writing. In the case of serpent, it's ideal to configure an editor that saves contracts for serpent using the suffix '.se in Python' and save '.Sol in order to guarantee the solidity. Don't use live refresh while you're working on the frontend of HTML since the testing of their contracts is not completed.

Setup of AlethZero

There is a lot of advancement taking place within IDE MIX, and even while there are many options in place the focus will be on the front-end, using this client AlethZero development that comes with the JavaScript console with a compiler as well as tools for peeking that look into the status that the blockchain is in.

Our tests will be run on AlethZero's single chain that is not connected to the networks. Development of contract on the testnet should be reserved for contracts which are to be shared with others. When alethzero runs with this configuration, additional users are able to join the chain as long as they have the same name and then connect via the connect-to-peer mode.

The first DApp

We will assist you in creating your first contract template although many details will be presumed for you to build your final product quickly. Let's take a look at Web applications that are decentralized.

Basics

The decentralized web , also known as "web3" is made possible by Ethereum. The main difference between 'web2' and" can be seen in the fact that Ethereum does not make use of web servers. There is also there is no middleman to waste commissions or steal information , and there isn't any DDoS.

Decentralized apps have an interface created using HTML as well as a backend, which is known as the database.

If you're into the bootstrap framework it is possible to use the framework as DApp's frontend has the ability to access the network completely and CDN's are also available. The frontend HTML writing of DApp is similar to website development, and the conversion from web 3 from web 2 is typically not significant.

If you're a Ruby or meteor lover you'll love the dynamic programming contained in it thanks to using callback function. Another benefit is that every DApp recognises the pseudonymous identity of each user, due to Ethereum's cryptographic principles that work. Simply put as a user you don't need to sign up for an account in order to access the DApps Think of it as an automatic setting.

Installing the client

Most stable s master build , as the client isn't as stable. We'll be using the AlethZero as well as it's C++ implementation for developers. You must install the master that has the most recent features.

Download the Windows as well as OSX binaries, and then follow particular instructions specific to Ubuntu that can be found on the web.

AlethZero Overview

After you've started AlethZero You are expected to be able to perform similar things to this:

The appearance of the interface is usually dependent to the screen's resolution. display.

In the middle of the screen is a browser that is known as the WebKit. The WebKit allows browsing from the WebKit and it's similar to the standard browser. Other panes contain information about technical aspects and debugging. This is helpful for the developer as well as other users. This design is different from the 'Mist' look that can be described as an Ethereum browser. After Ethereum is released, it will to sport a totally different design as illustrated below.

It is possible to rearrange the screen as you like. Panels can be moved and dropped over one another to create stacks.

Options

You can develop financial applications as well as games, social networks and even gambling applications on Ethereum since it's an open source programming platform. We'll create a basic contract that acts as a bank, however, it's an open ledger and auditable by all the world. The 10,000 tokens that will be utilized, and because it won't be enjoyable to keep the entire tokens for ourselves the method has to be developed to distribute the tokens out to our friends.

It is a straightforward method to issue our money. With web2, it will be impossible to create this kind of application in MySQL as well as PHP as users would be able to trust your accounts.

Contract

The contract acts as the backend and utilizes the Solidity language. Other contract language options exist that could be used to construct Ethereum's backend Serpent as well as LLL. Solidify is the one to use since it is the official language accepted by ETH-DEV.

We will create two things to help build our small-sized bank.

1.) Create accounts with tokens to allow us to begin.

2.) To move tokens around, we'll create a function to send tokens

Let's get started and then.

```
contract metaCoin{
mapping (address = balances in uint
function metaCoin function() {function
metaCoin()
balances[msg.sender] = 1000
|}
function sendCoin(address receiver and
uint amount)
Returns (bool sufficient) {Returns (bool
sufficient)
if (balances[msg.sender] < amount) return
false;
balances[msg.sender]-= amount;
balances[receiver] + the amount
True return
|}
```

Do not fret if you are unable to comprehend the code in the above article it's not as complex as it appears. Contracts are divided into different methods. The first is known as metaCoin. It's a specific constructor which defines the basic status of a storage contract. Constructor functions are known as contracts. This initialization code will executed once following the signing of the contract.

The contract code is akin and is the component that is a permanent part of the Ethereum network. In our case there is an element that checks that the sender's balance is sufficient enough to warrant that, if it is found to be true the transfer of the token will be made to a different account.

In detail,

mapping address = uint) balances

This code makes a mapping to a storage which allows the program to write data into the storage within the contracts. In this program the mapping is specified for key value pairs which comprise the type address as well as the uint, which is defined as balances. This is where you can find the balances on coins. The two types of data we have dealt with are the uint as well as the address.

function metaCoin function() {function metaCoin()

balances[msg.sender] = 10000;

|}

This is the process of contract initiation that is run only at once, and it will do many things. It first looks up public addresses with msg.sender which is the name of the person who sent the transaction that is, in this case, you. this instance. Second, it accesses accounts storage by with the help of mapping balances.

Let's examine the'sendCoin' function, which is executed when the contract is activated. It is the only executable function. Two arguments are present to the function: receiver and amount. Receiver is 160-bit , public address while amount represents tokens that are sent to the receiver.

The balance will be assessed at the beginning of each line. If it's lower than the tokens that are being sent, the second code isn't performed. If the balance is more than enough, there will be a false conditional evaluation and the amount will be subtracted by the two lines from the balance: balances[msg.sender]-= amount;,and the balance of the recipient's account is added.Balances[receiver] += amount;.

Now you can use a function that transfers tokens between accounts.

There are additional aspects that you must know about DApp which can be found in more in-depth books. There is still a lot to be learned about how to connect with Storage, JavaScript API 1& 2, and 3, contracts that send transactions variables, contract interaction with Event Logging Gas & Gas Price and how to utilize Mix.

Chapter 15: Trading Ethereum

If you're the adventurous kind, you might want to think about selling your Ether. It isn't the most popular method to earn money since it puts the investor at risk to greater risk than just purchasing and holding. Because of the high degree of risk, trading isn't the most suitable option for everyone.

Trading can be more complex than simply purchasing and selling. The trader is in fact betting on whether the price will rise or decrease. As an example, one could be able to conclude that Ether will increase in value in comparison to the US dollar or you could decide to bet on a different cryptocurrency.

If you think that Ether will increase of value then make a bet in an expectation that Ether's price will rise. But, if you are of the opinion that Ether's value will decline the bet will be placed in line with that.

There are numerous advantages of trading. Since you're not purchasing the

Ether the Ether, you don't need to contribute the entirety of the investment you're contemplating. You're able to expose yourself to higher profits, without losing everything you invested.

On the other hand traders who decide to remain in the market for extended durations are more susceptible to losses than investors would be.

In order to be successful when trading, you should be well-versed in market trends in terms of its history, as well as its patterns. You need a certain attitude to succeed in this kind of profitable business.

What Does It?

Trading is a difficult subject and can be difficult to grasp. For those who don't have a solid understanding of the technical aspects will find themselves having a difficult time understanding the basics of trading. It's also a stress-inducing way to earn money, people who are inclined to be overwhelmed will have a tough getting through the essential tenets of trading success.

There are some traits that you can observe when you look for successful traders. Check these out to see how your personality aligns with these aspects of psychologicality that make successful traders.

There isn't the usual excitement about Cryptocurrency: Many people, who hear of the potential money to be made in the digital realm, are eager and eager to pour their money into the pot. However, the trader will be able to hear about these stories, increase his curiosity but will want to do some more research before making the decision.

Willing to follow an authority If you do not have sufficient understanding of cryptocurrency, you'll be more likely to seek out experts who have already experienced some success with trading. The traders are impulsive, but they aren't impulsive enough to follow every story. However, should they meet an expert with a solid track record of showing them how to do it then they'll usually take their cues from them.

You do not have to do Anything: The market for cryptocurrency is always changing. The fluctuation of the market is what makes it extremely uncertain. Sometimes the only way an investor can earn a profit is to not do anything. For the majority of newbies, they believe they must perform something in order to succeed but they don't. They have to be selling or buying to keep their attention. The traders can tell the best time to not do anything, and keep an eye on what the market does.

Trading is a very emotionally-charged venture. The rapid pace of trading and sudden market shifts and the tumultuous highs and lows can be enough to drive the most rational person to the point of madness. Anyone who decides to trade must stand still in the midst that is chaotic, study the constantly changing flow of data and discover the tiny kernels of useful information to use as a guideline for how they go in or out of markets. They will need to withstand the FOMO (fear of being left out) and make the right decisions, that aren't easily influenced by the endless stories and rumors they are likely to be hearing.

This isn't for the faint-hearted If you are able to stand up to this sort level of stress could be a lucrative way to earn a substantial profit from your investment, while investing only a small portion of the total cost.

What Does It Mean

The process of trading any cryptocurrency is like what you do when participating in the Forex Trade. In contrast to investing, where you purchase at a specific price and then wait for it to rise it is trading to buy an alternative currency.

With Forex trade (foreign exchange) it is essentially the exchange of one currency for another in the expectation that the chosen currency will have more value in the near future. That means that when you trade Ether it is considering different currencies to determine which will be more powerful in the coming days or weeks.

At present, the two most prominent digital currencies in the world in the world is Bitcoin and Ethereum There isn't any rivalry between the two. Therefore, Ether traders Ether traders will likely be looking at fiat currencies in order to make gains. The problem to Ether trader is Ethereum isn't like any other cryptocurrency available on the market. It wasn't created to replace any other currency. It isn't used by anyone it as they would currencies. Therefore, the value of it in relation to dollars and cents could be extremely difficult and uncertain.

The majority of the risk lies in predicting the direction that occur in Ether against other currencies that are fiat. There is a variable that is not known that traders' minds must adjust for. Everything that is associated with Ether cannot be described in black and white so it makes each choice more prone to risk than other more well-defined altcoins.

Strategies for Trading

There are many trading strategies that be used when working with Ethereum

However, you must understand the fundamental steps to execute the trade on an exchange.

After a trader has determined which area he believes the best chance of earning a profit then he must make an order on the exchange. If he is looking to "buy" in this market, it puts what's called the open market position. If he is looking to "sell" the market, he takes the position of a close. The pending orders are the ones which he puts in place to take effect after certain conditions are met. Whatever the case, every type of order has to be utilized in the process of purchasing as well as selling Ether.

It is highly advised that traders make use of limit or stop orders to avoid the risk of being more exposed. Instead of purchasing or selling through the exchange instead, you can put a market-order instead. When it is placed, it is executed at the cost that will be incurred the moment it is made. If, for instance you place an order to purchase the item, you should also decide on your limit simultaneously.

The process is slightly different when you make an order to sell. When the market buy order is priced at the price for the sale, the sale order is filled with the bid price.

When you place these orders You won't have to be worried about price, however you'll need to specify what you're buying. The primary concern at this point, is what's known as "slippage," which is the difference between the amount you anticipate paying and the amount you could be charged. Although, in many circumstances, this might not be significant but due to the continuous fluctuation in prices, it may turn out to be significant. This typically happens when there is a great deal of activity taking place on the exchange simultaneously. As long as you know the possibility of this happening and you've prepared for the possibility, your trade will be successful.

Margin Trading

Margin trading is founded on a basic concept. You make the same kind of trades that you would on the standard trading platform, however you're not staking the entire order. You are charged a proportion of the cost of the transaction (the margin) and the remainder is provided by an additional entity, usually an broker house or an exchange.

If you invest $3,000 then that's your margin. The total value of your order will be contingent on the margin percentage. For instance, if you have a margin of 1:1 that means your investment of $3000 will constitute 1/3 of the total investment. This is a fantastic method to increase risk of profit, however it also puts you at risk of more losses.

This kind of investing comes with the greatest risk. Since the market is always moving it is possible to make a profit at any moment. The investor has the possibility of earning much higher profit in a short amount of time, however the risk exposure is far from the norm. Be aware that the greater leverage, the less risk, therefore 1:2 leverage will be far superior to 1:5. Some exchanges can offer leverages as low as 1:20, however it would be best to remain cautious of these opportunities. The potential for huge gains are real but the risks of losing are so high that it could be a sign that it's not worth it. Remember that you'll be required to make payments for interest on your loan amount. The interest is typically added on a daily basis which means that it will accumulate quickly. This is the reason traders are very involved on the market. They are in and out in a matter of minutes. The longer they stay the longer, the higher interest they will must pay, which can easily drain the profits before they have a chance to meet them.

If you're a margin trading persona and would like to try it, here are a few tips that can help you improve your odds of success.

Make your purchase after you have seen the price has seen an impressive drop.

You can close your position only if you earn 50 percent or more.

You can close your position once you've experienced a loss of less than 20 percent.

We've said it before: trading, particularly margin trading, isn't for the weak of heart. The key to success lies in the micro-profits and not the huge haul. Traders are satisfied with the accumulation of small profits that accumulate over time instead of one huge trade that could bring them the ultimate win for every single thing they've done.

Day Trading

Day Trading is a different high-intensity profitable business. It's about trading and then leaving in the exact same day. The most important rule for day traders is that you should never let your money sit on the market while you're asleep.

It requires a lot of skill to be a successful day trader. It takes a significant amount of time and energy to keep track of the market and identify what are your exit and entry places. Many people will sit on their screens for hours as they analyze their current situation and take strategic choices. This kind of trading is best for people who can be engaged in the market right at the beginning. If you're looking to make a more passive income Day trading is not the right choice the right choice for you.

Certain people have come up with a compromise to the day trading approach. Since it's going in and out one day. They opt to trade regularly however, not all day. They could choose each week a day to engage in active trading and the remainder of their time may be devoted to other activities.

Trading is among the most intricate ways of investing that you can discover, particularly on the market for cryptocurrency. The people who are most successful have mastered the art over time. Many who invest in trading cryptocurrency were investors in different markets before they brought their expertise to the new investment tool. If you decide to explore trading Ether ensure that you research thoroughly from reliable sources before beginning.

Chapter 16: Should I Invest In Ethereum?

There is no doubt that Ethereum is among the most promising investments in technology over the past few years. It has proven lucrative and has been increasing exponentially. Since the Ethereum's creation in 2016 this cryptocurrency has risen to over 1000 percent over its original value. It's literally transformed people earning minimum wage into millionaires in a relatively short period of. The growth of Ethereum is also increasing. If you had invested in February, you would have more than quadrupled in the last year.

Based on Ethereum's present pace of development, the coin could be worth 500 dollars within the next few years. The analysts believe that the future could be at the end of 2017. The long-term forecasts put the value of Ethereum up to thousands of dollars. If you're looking to invest in the cryptocurrency, it is best to take action now before it becomes more valuable.

As we've discussed in a prior chapter Ethereum is, in contrast to Bitcoin is much more than an electronic currency. It's a system that lets applications operate within it which leads to a variety of possibilities and options for services supplied by merchants.

What is it that can you do to make Ethereum an investment that is worthwhile? Here are three top reasons:

More applications compared to Bitcoin

Overall, digital currency has numerous benefits over traditional ones. They are safe, efficient and speedy. They also are immune to inf ation. One thing that differentiates cryptocurrency from traditional currencies is blockchain technology, which is their the technology that is its keystone.

Bitcoin became the first digital currency that was based on the blockchain network. The transactions that are processed on Bitcoin's Bitcoin blockchain are recorded in blocks that are distinct. These blocks are then merged with the others in the chain. The security of the block assures its security on the blockchain. It also enhances the security for the money itself. Blockchain is generally thought to be the most secure ledger. Digitally, it is.

What that makes Ethereum more intriguing is the fact that it uses an extremely advanced blockchain technology than the one that Bitcoin uses. Ethereum is coded in the Turing-complete programming language. To be considered as Turing-complete computers should be capable of running any algorithm. Due to this the coding language, all scripts is able to be executed within Ethereum. The blockchain utilized by Ethereum can track transactions much faster that Bitcoin can. Bitcoin completes transactions within 20 minutes. Ethereum can complete the transaction in only 12 seconds.

A more efficient network means more efficient applications and quicker processing. This is why the Ethereum blockchain is regarded as the most efficient network to support any kind of program or business. It is able to solve issues with speed and precision that can be matched by no other cryptocurrency.

Imagine an ordinary Honda car that has an engine from the supercar. In the assumption that there aren't any technical issues and the engine is running flawlessly within the vehicle. The regular car is now more efficiently using identical chassis. This is what the Turing-complete language does on the blockchain that is used by Ethereum.

Due to the speed of its operation, the precision, and the flexibility of Ethereum's blockchain, hundreds of major companies have poured their money in it, and each expects to reap the benefits of the system. This leads us to the second reason as to why Ethereum is a wise investment. It is backed by major corporations.

Fortune 500 Companies Back Ethereum

The Enterprise Ethereum Alliance or the EEA is an important element of proof that will ensure the durability of the cryptocurrency. It was established in February 2017 and is comprised of a variety of Fortune 500 companies. They collaborated to improve the technology that powers the Ethereum network with the objective of the cryptocurrency being integrated into their own business. The group is comprised of major corporations like Microsoft, Intel, BP, J.P. Morgan, and Thomson Reuters.

These companies take their risks seriously. They're at the top of their game due to their great executive decisions and continually create business models.

One of the causes behind Ethereum rapid growth is that the EEA announced the creation of the group to the general public. The businesses that comprise the EEA recognize that there are a variety of reasons to consider Ethereum to be attractive. The cryptocurrency's blockchain is acknowledged for its efficiency and speed. The Ethereum network is able to handle smart contracts.

Smart contracts will be discussed in depth in a different chapter. However, with an example, suppose that you've reached this agreement with your cousin: If you strike him, he will hit you and hit you back. Now imagine the interaction takes place in the code world. The smart contract decides on the offence (you hit your friend) and then executes the agreed-upon react (your neighbor hitting you). Smart contracts make things run smoothly. They are the foundation of the blockchain.

Ethereum was created with the intention of allowing smart contracts to operate efficiently. Prior to when Buterin developed Ethereum the platform, he had been involved in blockchain as well as Bitcoin. The genius developer discovered that Bitcoin cannot process smart contracts, which is a major flaw.

The Ethereum Virtual Machine or the EVM is located by the Ethereum network. It's responsible for processing smart contracts. It also takes decisions and determines the appropriate charges. The EVM could be a great tool for improve the efficiency of existing business processes and improving efficiency of business processes. It doesn't matter what the process is - demand, payment transactions, payment, etc. A prompt and timely reaction is provided for each step that is taken.

Efficiency can translate into money. This is what the EEA trend-setters view it. These CEOs from firms are well-known for their accuracy in predicting economic developments. They know that stepping in early in Ethereum can help them to position their businesses for the huge growth potential that is expected from the crypto. Investors are also taking note of this. A few of these investors have started to buy stocks from these early mover with the hopes of reaping the rewards that will be realized in the longer term. It is inevitable that Ethereum will become a component of all businesses.

Financial Institutions are incorporating Ethereum

Cryptocurrencies are being embraced by the general population. Ethereum is well-positioned and has the most potential to other currencies. This adoption, ironically, began with institutions that are expected to fall victimized by the digital currency, namely banks.

Bitcoin is believed to be the bank's adversary. It's known to be a danger to the current financial system, which is the reason that gives Bitcoin significance in the eyes of some people.

Ethereum is a better choice for banks as it permits banks to thrive in the world of digital. Bank of America has taken the first step towards working using its Ethereum blockchain. It has created an application built on Ethereum which helps customers to ensure the security of transactions. The application was created with Microsoft's assistance. The objective is to accelerate Ethereum technology's acceptance into the mainstream financial market.

The application converts customer data into blockchain packets. Only those who are part of the transaction in the current transaction are able to access the data. The risk of information leaks is reduced since the information isn't transmitted via email. This removes many privacy issues.

Ethereum is thought to have incredible capabilities in the field of security financial transactions, and this application is proof of that. Microsoft as well as Bank of America hope this application will be more than an information security tool. It needs to earn public's trust and be the first in this area. Soon, the everyday users will be interacting with similar applications made up of Ethereum. The adoption of the latest technology will take time and consumers will not even be aware that they already use it.

When people buy new versions of MacBooks or iPhones, Apple is guiding users to use the latest technology behind the brand new product. Similar is the case for Ethereum. Ethereum blockchain. The big companies will gradually integrate the blockchain into their operations and then pass it on to clients and customers. Ethereum is believed as a digital currency by many companies and is believed to offer significant advantages over other currencies.

At present, Ethereum alone is being taken over by big organizations, as they've had more than an entire decade to work out using Bitcoin. This is ensuring that

Ethereum's financial viability over the long run.

Chapter 17: Ethereum And Other Cryptocurrencies

Ethereum is currently the second most popular cryptocurrency in the world, just behind Bitcoin. Some people even believe the digital currency is poised to eventually surpass BTC.

But, it's a good investment decision to not put all your eggs into one basket. The old method of diversification remains relevant to the cryptocurrency you hold. It is therefore essential to be aware of other cryptocurrency with potential.

Bitcoin Cash (BCH)

A brief period of time in the final period of the 2017 quarter, discord was brewing among those who have adopted Bitcoin regarding the technical limitations of Bitcoin. This led in a fork within the blockchain. This led in the launch to Bitcoin Cash (BCH).

A group of disappointed cryptocurrency miners decided to split the token using a different software with the aim of securing the cryptocurrency's growth. Since its debut the token has earned its place in the list of top cryptocurrency without having to take over Bitcoin in terms of the buzz, value, or use. BCH is estimated to have a market cap of over $28 billion.

Monero (XMR)

A different open-source, decentralized digital currency is Monero. The Monero currency, which is private and not traceable, was launched in the first quarter of 2014.

In a brief time it has managed to spark an interest in that cryptography group. It has successfully convinced numerous people to become investors. The development of Monero is entirely donated and driven by community.

The digital currency was designed to be scalable and decentralization. It provides additional security by using a unique algorithm for cryptography called "ring signatures". In this method, the cryptographic signatures of a group are presented (including the one that is the actual participant) However, as each signature appears to be legitimate, the authentic one is not able to be recognized.

Cardano (ADA)

In the beginning, Cardano is a platform which is used to transfer digital currencies. It facilitates storage and the transfer of value via the ADA token. Similar to Ether the Cardano network Cardano is designed to run without centralized control of applications on the blockchain. It was designed by a co-founder and co-founder of the Ethereum Network, Charles Hopkins. It is also considered to be one of the Ethereum of Japan since around 90 percent of its ICOs initiated by Japanese. The network is currently managed by a team of academics and researchers who are experts in blockchain-related applications. The market capitalization of Cardano is believed to be approximately $16 billion.

Ripple (XRP)

Ripple is a crypto currency that was launched in 2012. It has a current market cap at $1.26 billion. It is a global settlement system that is real-time that allows immediate, precise inexpensive international transactions. It allows banks as well as similar financial institutions to manage payments across borders in close-to-real-time (if not in real-time) with a low cost.

Ripple's model does not require mining with GPU as well as CPUs, so it reduces dependence on computational capacity , and also reduces the latency of networks. The developers of Ripple believe that dispensing price is a successful method of rewarding certain actions and therefore they're planning to distribute the tokens via deals and investments by institutions.

Dash (DASH)

A more obscure alternative to Bitcoin was introduced during the beginning of 2014. Dash Dash (also called Darkcoin) was developed and was developed by Evan Duffield. It was in the year 2015 that Darkcoin changed its name to Dash which translates to Digital Cash. It is mined with either a GPU as well as a CPU.

The change in its name didn't affect any of its features such as InstantX or Darksend

The Dash service continues to provide anonymity since it runs using a mastercode decentralized web which makes transactions virtually impossible to trace. Dash is currently being used by many people across the world.

Zcash (ZEC)

In the third trimester of the year, an open source and decentralized digital currency was introduced. It was called Zcash. It was described as the"https of money in comparison to Bitcoin was regarded as the http. Zcash offers privacy and transparency for all transactions.

Therefore, Zcash still claims to offer extra security as each transaction is tracked and recorded through a blockchain, however details such as the recipient, sender as well as the amount, are not listed. "Shielded" transaction are provided by this digital currency that allow the data to be protected using the latest techniques for cryptography. The cryptographic method Zk-SNARK is a zero-knowledge proof technique that was created by the same group of people who developed Zcash.

Conclusion

We've reached the conclusion of the book. Thanks for your time reading, and congrats for having read to the very end.

In this section of the guide, you're now ready to begin using cryptocurrency because you are knowledgeable about them and how they work and how you can invest and trade in them in a safe way.

As you've seen that cryptocurrency is safe and secure. They are also promising too. They address a myriad of problems that are in traditional banking, and even when using a digital platform. I want to remind you of the fact that cryptocurrency like Bitcoin make use of a complicated blockchain technology to ensure that funds are not duplicated or "double spent"--and If you study the book thoroughly you will understand why , or at least , know more than the majority of people. This is just one among many advantages of cryptocurrency.

If you've been wondering what all the fuss and commotion regarding cryptos was all about prior to this, now you're aware!